About City & Guilds

City & Guilds is the UK's leading provider of vocational qualifications, offering over 500 awards across a wide range of industries, and progressing from entry level to the highest levels of professional achievement. With over 8500 centres in 100 countries, City & Guilds is recognised by employers worldwide for providing qualifications that offer proof of the skills they need to get the job done.

Equal opportunities

City & Guilds fully supports the principle of equal opportunities and we are committed to satisfying this in all our activities and published material. A copy of our equal opportunities policy statement is available on the City & Guilds website.

Copyright

First edition 2014

ISBN 978-0-85193-280-4

Publisher: Charlie Evans

Development Editor: Hannah Cooper

Production Editor: Fiona Freel

Picture Research: Katherine Hodges

Project Management and Editorial Series Team: Vicky Butt, Anna Clark, Kay Coleman, Jo Kemp, Karen Hemingway, Jon Ingoldby, Caroline Low, Joan Miller, Shirley Wakley

Cover design by Select Typesetters Ltd

Text design by Design Deluxe, Bath

Indexed by Indexing Specialists (UK) Ltd

Illustrations by Saxon Graphics Ltd and Ann Paganuzzi

Typeset by Saxon Graphics Ltd, Derby

Printed in the UK by Cambrian Printers Ltd

Publications

For information about or to order City & Guilds support materials, contact 0844 534 0000 or centresupport@cityandguilds.com. You can find more information about the materials we have available at www.cityandguilds.com/publications.

Every effort has been made to ensure that the information contained in this publication is true and correct at the time of going to press. However, City & Guilds' products and services are subject to continuous development and improvement and the right is reserved to change products and services from time to time. City & Guilds cannot accept liability for loss or damage arising from the use of information in this publication.

City & Guilds
1 Giltspur Street
London EC1A 9DD

T 0844 543 0033
www.cityandguilds.com
publishingfeedback@cityandguilds.com

THE CITY & GUILDS TEXTBOOK

LEVEL 3 NVQ DIPLOMA IN
ELECTROTECHNICAL TECHNOLOGY 2357
UNITS 307-30

THE CITY & GUILDS TEXTBOOK

LEVEL 3 NVQ DIPLOMA IN
ELECTROTECHNICAL TECHNOLOGY 2357

UNITS 307–308

HOWARD CAREY AND TREVOR PICKARD
SERIES EDITOR: PETER TANNER

ACKNOWLEDGEMENTS

City & Guilds would like to thank sincerely the following:

For invaluable knowledge and expertise

Geoff Cronshaw, *CEng*, *FIET*, The Institution of Engineering and Technology, Technical Reviewer

Richard Woodcock, City & Guilds, Technical Reviewer and Contributor

For supplying pictures for the front and back covers

Jules Selmes

For their help with photoshoots

Jules Selmes and Adam Giles (photographer and assistant); Andy Jeffery, Ben King and students from Oaklands College, St Albans; Jaylec Electrical; Andrew Hay-Ellis, James L Deans and the staff at Trade Skills 4 U, Crawley.

Picture credits

Every effort has been made to acknowledge all copyright holders as below and the publishers will, if notified, correct any errors in future editions.

Alamy p157; **BSI** p36, p40, p46, p69, p88, p89, p91, p97, p110, p111, p113, p159 (Permission to reproduce extracts from British Standards is granted by BSI Standards Limited (BSI). No other use of this material is permitted. British Standards can be obtained in PDF or hard copy formats from the BSI online shop: www.bsigroup.com/Shop or by contacting BSI Customer Services for hard copies only: Tel: +44 (0)20 8996 9001, Email: cservices@ bsigroup.com); **Hager** p109; **HSE** p3, p7, p12, p50, p150, p153, p191; **IET** p5, p14, p16, p17, p18, p19, p21, p22, p23, p26, p30, p32, p38, p41, p54, p56, p58, p59, p74, p76, p78, p80, p83, p85, p87, p89, p90, p93, p94, p95, p100, p104, p106, p108, p109, p116, p117, p118, p119, p121, p212, p213; **Insulated Tools Ltd** p155 (www.insulatedtoolsgroup.com); **Kewtech** p47, p48, p74, p193, p194, p195, p204; **Martindale Electric Company Ltd.** p190, p206; **Megger Ltd.** p50, p53, p82; **Shutterstock** p1, p6, p144, p155, p181, p185; **test-meter.co.uk** p5; **Wylex Electrium** p100.

Text permissions

For kind permission to use text extracts:

Crown copyright. Contains public sector information published by the Health and Safety Executive and licensed under the Open Government Licence v.1.0 and includes extracts from the following HSE publications: 'Dangerous Substances and Explosive Atmosphere Regulations' page 185; 'Electricity at Work Regulations' pages 3, 4, 7, 26, 149, 159.

Permission to reproduce extracts from BS 7671:2008 is granted by the Institution of Engineering and Technology (IET) and BSI Standards Limited (BSI). No other use of this material is permitted. Pages 149, 159, 173, 198, 214. BS 7671: 2008 Incorporating Amendment No 1: 2011 can be purchased in hardcopy format only from the IET website http://electrical.theiet.org/ and the BSI online shop: http://shop.bsigroup.com

From the authors

Howard Carey: I would like to thank Shirley Wakley for her professionalism, patience and guidance in working with my material.

Trevor Pickard: I need to express my sincere thanks to my wife, Sue, whose command of the English language has always been better than mine and without whose help the job of the copyeditor would have been that much harder!

Peter Tanner: Many thanks to my wife, Gillian, and daughters, Rebecca and Lucy, for their incredible patience; also to Jim Brooker, my brother-in-law, for showing me the importance of carrying out the correct safe isolation procedure on more than one occasion!

CONTENTS

ABOUT THE AUTHORS

Howard Carey *MIET, LCGI, CertEd*

My career in electrical installation started when I left school at the age of 16. A five-year apprenticeship with a small electrical engineering company allowed me to gain industrial, commercial, agricultural and domestic knowledge of electrical installation and maintenance. During this period I attended New College Durham on a day release course. I also attended evening classes and attained a City & Guilds 'C cert', a foreman electrician qualification. I have gained many other electrical qualifications, but this was by far my favourite.

After a further five years of experience in the contracting industry, I took a job as an instructional officer in the Civil Service. After six years in this position I became a freelance lecturer, assessor, test engineer and consultant, which I am continuing to enjoy now, around 20 years later.

I have achieved assessor awards and teaching qualifications including City & Guilds and a Certificate in Education. My ultimate achievement was in November 1999 when I became a member of the courses lecturing team with The Institution of Engineering and Technology (IET).

My working experience has involved me teaching in England, Scotland, Wales and Cyprus, on a range of courses including inspection and testing, wiring regulations, health and safety and industrial systems.

My advice to any apprentice or learner is to work hard at college and keep in touch with the real world. Electrical installation work and regulations are ever-changing topics and there is always something new to learn. Remember to dedicate a little time to furthering your qualifications and continuing your professional development.

Trevor Pickard *IEng, MIET*

I am an electrical engineering consultant and my interest in all things electrical started when I was quite young. I always had a battery powered model under construction or an electrical motor or some piece of electrical equipment in various stages of being taken apart to see how they operated. Looking back, some of my activities with mains electricity would certainly be considered as unacceptable today!

Upon leaving school in 1966 I commenced work with an electricity distribution company, Midlands Electricity Board (MEB) and after serving a student apprenticeship I held a series of engineering positions. I have never tired of my involvement with electrical engineering and was very fortunate to have had a varied and interesting career in the engineering department of Midlands Electricity and embraced its various changes through privatisation and subsequent acquisition. I held posts in Design, Safety, Production Engineering, as Production Manager of a large urban-based operational division, as General Manager of the Repair and Restoration department, and as General Manager of the Primary Network department (33kV–132kV).

My interest in electrical engineering has extended beyond the '9–5 job' and I have had the opportunity to become involved in the writing of standards in the domestic, European and international arena with BSI, CENELEC and IEC and have for many years lectured for the Institution of Engineering and Technology.

Electricity is with us in almost every aspect of modern life and for those who are just starting their career in this field I would say keep an open mind, be safety conscious in how you carry out your work and who knows where your studies will take you.

Peter Tanner *MIET, LCGI*

Series Editor

I started in the industry while still at school, chasing walls for my brother-in-law for a bit of pocket-money. This taught me quickly that if I took a career in the industry I needed to progress as fast as I could.

Jobs in the industry were few and far between when I left school so after a spell in the armed forces, I gained a place as a sponsored trainee on the CITB training scheme. I attended block release at Guildford Technical College where the CITB would find me work experience with 'local' employers. My first and only work experience placement was with a computer installation company located over twenty miles away so I had to cycle there every morning but I was desperate to learn and enjoyed my work.

Computer installations were very different in those days. Computers filled large rooms and needed massive armoured supply cables, so the range of work I experienced was vast, from data cabling, to all types of containment systems and low voltage systems.

In the second year of my apprenticeship I found employment with a company where most of my work centred around the London area. The work was varied, from lift systems in well-known high-rise buildings to lightning protection on the sides of even higher ones!

On completion of my apprenticeship I worked for a short time as an intruder alarm installer, mainly in domestic dwellings, a role where client relationships and handling information is very important.

Following this I began work with a company where I was involved in shop-fitting and restaurant and pub refurbishments. It wasn't long before I was managing jobs and gaining further qualifications through professional development. I was later seconded to the Property Services Agency, designing major installations within some of the most well-known buildings in the UK.

A career-changing accident took me into teaching where I truly found the rewards the industry has to offer. Seeing young trainees maturing into qualified electricians is a worthwhile experience. On many occasions I see many of my old trainees when they attend further training and update courses. Seeing their successes makes it all worthwhile.

I have worked with City & Guilds for over twenty years and represent them on a variety of industry committees such as JPEL64, which is responsible for the production of BS 7671. I am passionate about using my vast experience in the industry to maintain the high standards the industry expects.

HOW TO USE THIS TEXTBOOK

Welcome to your City & Guilds Level 3 NVQ Diploma in Electrotechnical Technology textbook. It is designed to guide you through your Level 3 qualification and be a useful reference for you throughout your career.

Each chapter covers a unit from the 2357 Level 3 qualification. Each chapter covers everything you will need to understand in order to complete your written assignments or online tests and prepare for your practical assessments. Across some learning outcomes in the 2357 units there is some natural revisiting of knowledge and skills in different contexts, which the content in this book also reflects, for practical use and reference.

Throughout this textbook you will see the following features:

KEY POINT

The user of the instrument should always check to ensure the instrument is within the calibration period before using it.

KEY POINT These are particularly useful points that may assist you in revision for your tests or help you remember something important.

SELV

Separated extra-low voltage circuit

DEFINITIONS Words in bold in the text are explained in the margin to aid your understanding. They also appear in the glossary at the back of the book.

ACTIVITY

Short circuits may often blow clear. What is meant by this?

ACTIVITY These provide questions or suggested activities that help you learn and practise.

ASSESSMENT GUIDANCE

You will be expected to be seen checking test instruments before and after use.

ASSESSMENT GUIDANCE These highlight useful points that are helpful for your learning and assessment.

 SmartScreen Unit 308
Handout 20

SMARTSCREEN These provide reference to City & Guilds online learner and tutor resources, which you can access on SmartScreen.co.uk.

Assessment criteria

4.2 Describe how to identify supply voltages

ASSESSMENT CRITERIA These highlight the assessment criteria coverage through each unit, so you can easily link your learning to what you need to know or do for each learning outcome.

You will also see the following abbreviation in the running heads:
LO – learning outcome   **LO4**   **Diagnosing faults**

Where tables and forms in this book have been used directly from other publications such as the IET this has been noted, and the style reflects the original (with kind permission). Always make sure you use the latest information and forms.

Understanding principles, practices and legislation for the inspection, testing, commissioning and certification of electrotechnical systems and equipment in buildings, structure and the environment

Inspection, testing, commissioning and certification are very important stages of the safety process. Electricity can be very dangerous if the guidelines in the current edition of the Institution of Engineering and Technology (IET) Wiring Regulations (BS 7671) are not followed.

Electrical inspection and testing can involve some degree of risk but if the guidelines are followed, the risk can be minimised. Every year, the Health and Safety Executive investigates many cases of injury due to bad practice and poor workmanship. Electric shock, burns and fatalities can happen; however, if a member of a workforce is killed at work, the organisation could face a corporate manslaughter case. Employers, managers and employees can be fined or sent to prison for neglecting health and safety guidelines.

This unit covers the health and safety requirements, and includes the theory and testing skills required to become competent during the process of inspection, testing and commissioning of electrical installations.

LEARNING OUTCOMES

There are five learning outcomes to this unit. The learner will:

1 understand the principles, regulatory requirements and procedures for completing the safe isolation of an electrical circuit and complete electrical installations in preparation for inspection, testing and commissioning

2 understand the principles and regulatory requirements for inspecting, testing and commissioning electrical systems, equipment and components

3 understand the regulatory requirements and procedures for completing the inspection of electrical installations

4 understand the regulatory requirements and procedures for the safe testing and commissioning of electrical installations

5 understand the procedures and requirements for the completion of electrical installation certificates and related documentation.

This unit will be assessed by:

- online multiple-choice assessment (closed book)
- practical assessment to demonstrate knowledge of inspection and testing (open book)
- written examination with three scenario based questions and three general testing questions (closed book).

Understand the principles, regulatory requirements and procedures for completing the safe isolation of an electrical circuit and complete electrical installations in preparation for inspection, testing and commissioning

Assessment criteria

1.1 State the requirements of the Electricity at Work Regulations 1989 for the safe inspection of electrical systems and equipment, in terms of those carrying out the work and those using the building during the inspection

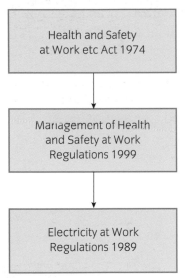

Statutory documents relating to inspection and testing

HEALTH AND SAFETY LEGAL REQUIREMENTS

The flowchart shows the three most relevant statutory (legal) requirements that apply during electrical inspection, testing and commissioning.

If an accident occurs in the workplace, the Health and Safety at Work etc Act 1974 (HSW Act) will be addressed initially, because this is the general standard for all workplace tasks – including electrical work. It puts the duty of care on both employer and employee to ensure the safety of all persons using the work premises.

If the accident is electrical in nature, other statutory requirements may also be relevant:

- *The Management of Health and Safety Regulations 1999* also apply to general workplace tasks and training, and state that every employer must make a suitable and sufficient assessment of the risks to which employees are exposed. Training must be provided for employees to be able to carry out the tasks that they are expected to do in the workplace.

- *The Electricity at Work Regulations 1989* are focused on electrical work and methods of safety. They must be followed to ensure that the electrical inspector, other trades and members of the public are not put at risk during the process of inspection and testing. Property and equipment must also be guarded against misuse and damage.

Importance of the legal requirements

It is vital to work in a safe way and that means following the legal requirements. If an accident occurs, for example, and an electrician gets blinded by arcing or by an explosion caused by a short circuit during testing, the following three questions could be asked in a court of law:

1 Did the employee take reasonable care when testing? (Regulation 7 of The Health and Safety at Work etc Act 1974)

2 Was the employee trained and capable of carrying out the task requested by the employer? (Regulation 13 of The Management of Health and Safety at Work Regulations 1999)

3 Was the electrician competent and could the work have been carried out with the circuit isolated? (Regulations 16 and 14 of Electricity at Work Regulations 1989)

Contravention of any of these regulations could result in the employee, such as the electrician, and the employer being found guilty of a crime, and incurring heavy fines or being sent to prison.

Electricity at Work Regulations

You are expected to know the Electricity at Work Regulations 1989 (EAWR) and to understand how these impose responsibilities on the **duty holder** (you), when inspecting, testing, commissioning and certificating a new circuit or installation.

Some important extracts from the EAWR 1989 are listed below.

PART II GENERAL – Electricity at Work Regulations 1989

Regulation 4 – Systems, work activities and protective equipment

(1) All systems shall at all times be of such construction as to prevent, so far as is reasonably practicable, danger.

Regulation 13 – Precautions for work on equipment made dead

Adequate precautions shall be taken to prevent electrical equipment, which has been made dead in order to prevent danger while work is carried out on or near that equipment, from becoming electrically charged during that work if danger may thereby arise.

Regulation 14 – Work on or near live conductors

No person shall be engaged in any work activity on or so near any live conductor (other than one suitably covered with insulating material so as to prevent danger) that danger may arise unless –

(a) it is unreasonable in all the circumstances for it to be dead; and

(b) it is reasonable in all the circumstances for him to be at work on or near it while it is live; and

(c) suitable precautions (including where necessary the provision of suitable protective equipment) are taken to prevent injury.

Regulation 15 – Working space, access and lighting

For the purposes of enabling injury to be prevented, adequate working space, adequate means of access, and adequate lighting shall be provided at all electrical equipment on which or near which work is being done in circumstances which may give rise to danger.

Regulation 16 – Persons to be competent to prevent danger and injury

No person shall be engaged in any work activity where technical knowledge or experience is necessary to prevent danger or, where appropriate, injury, unless he possesses such knowledge or experience, or is under such degree of supervision as may be appropriate having regard to the nature of the work.

KEY POINT

You can find additional information with in-depth explanations in the 'Memorandum of guidance on the Electricity at Work Regulations 1989'. This can be downloaded from the Health and Safety Executive website: www.hse.gov.uk

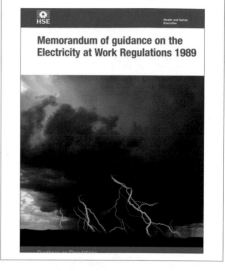

Duty holder

The person responsible for actions and matters that are within their control.

ACTIVITY

Read Regulation 29 of the Electricity at Work Regulations and discuss its importance regarding your legal rights.

SmartScreen Unit 307

Handout 3

How these regulations affect the inspection and testing process

The Electricity at Work Regulations 1989 affect the inspection and testing process as follows.

- Regulation 4(1) identifies the need to ensure that new electrical installations and circuits are subject to inspection, testing and commissioning before being put into service.

- Regulation 13 identifies the need for equipment to be 'made **dead**' during the process of initial verification when undertaking dead tests. Isolation procedures will need to be adopted (see pages 5 and 6 for isolation procedures).

- Regulation 14 identifies that working on or near **live** conductors will be necessary during initial verification, for example, in distribution boards and consumer units. Risk assessments, test equipment and safe working practices are covered in more detail on pages 10–12.

- Regulation 15 involves working space, **access** and lighting. Good design and installation methods will ensure that access to electrical equipment is in accordance with these requirements.

- Regulation 16 emphasises **competence** and the need to fully understand the task. There are many types of electrician – industrial, commercial and domestic, to name a few. When undertaking inspection, testing and commissioning, the duty holder must have the appropriate knowledge and experience of the system to work in a safe manner.

THE CORRECT PROCEDURE FOR SAFE ISOLATION

Isolation can be very complex due to the differing industrial, commercial and domestic working environments, some of which require experience and knowledge of the system processes.

This section deals with a basic practical procedure for the isolation and for the securing of isolation. It also looks at the reasons for safe isolation and the potential risks involved during the isolation process.

How to undertake a basic practical procedure for isolation

Gather together all of the equipment required for this task

You will need the following equipment:

- a voltage indicator which has been manufactured and maintained in accordance with Health and Safety Executive (HSE) Guidance Note GS38
- a proving unit compatible with the voltage indicator
- a lock and/or multi-lock system (there are many types of lock available)
- warning notices which identify the work being carried out
- relevant Personal Protective Equipment (PPE) that adheres to all site PPE rules.

The equipment shown in the photographs can be used to isolate various main switches and isolators. To isolate individual circuit breakers with suitable locks and locking aids, you should consult the manufacturer's guidance.

When working on or near electrical equipment and circuits, it is important to ensure that:

- the correct point of isolation is identified
- an appropriate means of isolation is used
- the supply cannot inadvertently be reinstated while the work is in progress
- caution notices are applied at the point(s) of isolation
- conductors are proved to be dead at the point of work before they are touched
- safety barriers are erected as appropriate when working in an area that is open to other people.

ACTIVITY

Check your approved voltage indicator for any damage and for compliance with GS38.

Voltage indicator   **Proving unit**

Lock-out facility – can be used with one or more locks

Lock-out devices for circuit breakers and RCBOs

Typical warning notices

Isolator locked and tagged (secured)

Circuit breaker locking device (not secured)

Carry out the practical isolation

The method of isolation is outlined below.

1 *Identify* – identify equipment or circuit to be worked on and point(s) of isolation.

2 *Isolate* – switch off, isolate and lock off (secure) equipment or circuit in an appropriate manner. Retain the key and post caution signs with details of work being carried out.

3 *Check* – check the condition of the voltage indicator leads and probes. Confirm that the voltage indicator is functioning correctly by using a proving unit.

4 *Test* – using voltage indicator, test the outgoing terminals of the isolation switch. Take precautions against adjacent live parts where necessary.

- During single-phase isolation there are three tests to be carried out:
L – N
L – E
N – E
(L = Line, N = Neutral, E = Earth)

- During three-phase isolation there are 10 possible tests (if the neutral is present):

L1 – N	L2 – N	L3 – N
L1 – E	L2 – E	L3 – E
L1 – L2	L1 – L3	L2 – L3
N – E		

(L = Line, N = Neutral, E = Earth)

5 *Prove* – using voltage indicator and proving unit, prove that the voltage indicator is still functioning correctly.

6 *Confirm* – confirm that the isolation is secure and the correct equipment has been isolated. This can be achieved by operating functional switching for the isolated circuit(s).

The relevant inspection and testing can now be carried out.

Reinstate the supply

When the 'dead' electrical work is completed, you must ensure that all electrical barriers and enclosures are in place and that it is safe to switch on the isolated circuit.

1 Remove the locking device and danger/warning signs.

2 Reinstate the supply.

3 Carry out system checks to ensure that the equipment is working correctly.

The requirements of the Electricity at Work Regulations 1989

When undertaking the correct procedure for isolation, you will need to abide by Regulation 13 and Regulation 14, as shown below.

PART II GENERAL – Electricity at Work Regulations 1989

Regulation 13 – Precautions for work on equipment made dead

Adequate precautions shall be taken to prevent electrical equipment, which has been made dead in order to prevent danger while work is carried out on or near that equipment, from becoming electrically charged during that work if danger may thereby arise.

Regulation 14 – Work on or near live conductors

No person shall be engaged in any work activity on or so near any live conductor (other than one suitably covered with insulating material so as to prevent danger) that danger may arise unless–

(a) it is unreasonable in all the circumstances for it to be dead;

How these regulations affect the inspection and testing process

The Electricity at Work Regulations 1989 affect the inspection and testing process as follows.

- Regulation 13 identifies the need for equipment to be 'made dead' during the process of initial verification when undertaking dead tests. Isolation procedures will be needed.

- Regulation 14 acknowledges that during isolation, an electrician may be working on or near live conductors.

ACTIVITY

Assemble the equipment required to isolate a circuit, and simulate isolation on various isolators and circuit breakers that are not connected to a supply. This can be done safely on a workbench or desk.

SAFE ISOLATION AND IMPLICATIONS

When you isolate an electricity supply, there will be disruption. So, careful planning should precede isolation of circuits.

Consider isolating a section of a nursing home where elderly residents live. You will need to consult the nursing home staff, to consider all the possible consequences of isolation and to prepare a procedure.

Electricity at work
safe working practices

HSE

ASSESSMENT GUIDANCE

You will be assessed by means of written or practical assignments and expected to answer questions relating to methods of isolation, testing and securing isolation.

Assessment criteria

1.3 State the implications of carrying out safe isolations to:

- other personnel
- customers/clients
- public
- building systems (loss of supply)

The following questions are useful:

1 How will the isolation affect the staff and other personnel?
For example, think about loss of power to lifts, heating and other essential systems.

2 How could the isolation affect the residents and clients?
For example, some residents may rely on oxygen, medical drips and ripple beds to aid circulation. These critical systems usually have battery back-up facilities for short durations.

3 How could the isolation affect the members of the public?
For example, fire alarms, nurse call systems, emergency lighting and other systems may stop working.

4 How can an isolation affect systems?
For example, IT programs and data systems could be affected; timing devices could be disrupted.

In this scenario, you must make the employers, employees, clients, residents and members of the public aware of the planned isolation. Alternative electrical back-up supplies may be required in the form of generators or uninterruptable power supply systems.

ISOLATION RISKS AND IMPLICATIONS

Before any isolation is carried out, you must assess the risks involved. This section deals with the practical implications and the risks involved during the isolation procedure, if risk assessment and method statements are not followed.

Who is at risk and why?

If isolation is *not* carried out safely, what are the possible risks when performing inspection and testing tasks?

Risks to you

Risks to you might include:

- shock – touching a line conductor, eg if isolation is not secure

- burns – resulting from touching a line conductor and earth, or arcing

- arcing – due to a short circuit between live conductors, or an earth fault between a line conductor and earth

- explosion – arcing in certain environmental conditions, eg in the presence of airborne dust particles or gases, may cause an explosion.

Assessment criteria

1.4 State the implications of not carrying out safe isolations to:
- self
- other personnel
- customers/clients
- public
- building systems (presence of supply)

ACTIVITY

Why is the N or E probe connected before the Line when carrying out this test?

Things that can *cause risk* to you include:

- inadequate information to enable safe or effective inspection and testing, i.e. no diagrams, legends or charts
- poor knowledge of the system you are working on (and so not meeting the competence requirements of Regulation 16 of EAWR)
- insufficient risk assessment
- inadequate test instruments (not manufactured or maintained to the standards of GS38).

Risks to other tradespersons, customers and clients

Risks to other tradespersons, customers and clients might include:

- switching off electrical circuits – for example, switching off a heating system might cause hypothermia; if lifts stop, people may be trapped.
- applying potentially dangerous test voltages and currents
- access to open distribution boards and consumer units
- loss of service or equipment, for example:
  - □ loss of essential supplies
  - □ loss of lights for access
  - □ loss of production.

Risks to members of the public

Risks to members of the public might include prolonged loss of essential power supply, causing problems, for example, with safety and evacuation systems, such as:

- fire alarms
- emergency exit and corridor lighting.

Note that, although safety services usually have back-up supplies such as batteries, these may only last for a few hours. Other safety or standby systems may have generator back-up, but this will also require isolation, leaving the building without any safety systems.

Risks to buildings and systems within buildings

Risks to buildings and systems within buildings might involve applying excessive voltages to sensitive electronic equipment, for example:

- computers and associated IT equipment
- residual current devices (RCDs) and residual current operated circuit breakers with integral overcurrent protection (RCBOs)
- heating controls
- surge protection devices.

There might also be risk of loss of data and communications systems.

ASSESSMENT GUIDANCE

You will be assessed by means of a written assignment and expected to answer questions relating to risks involved during the isolation, inspection and testing of buildings (closed-book questions).

ACTIVITY

1 Write down a list of the risks associated with isolation and the effects isolation can have on people, livestock, systems and buildings.

2 Who is at risk if inspection and testing is not carried out correctly?

3 What might happen if you need to switch off a socket outlet circuit in a hospital?

4 What must you do if you encounter a computer server that requires a permanent supply and you need to switch off the main supply to enable safe testing procedures?

ACTIVITY

Discuss with your class or colleagues, the dangers associated with working on or near live conductors.

RISK ASSESSMENTS, SAFE WORKING PROCEDURES AND EQUIPMENT

You should already be familiar with risk assessments, permits to work and method statements. This section focuses on how to apply these documents to the tasks involved in inspection and testing.

It is essential to have approved documentation, protective equipment, tools and test equipment. All of these must be supplied, maintained and updated (or replaced) as deemed necessary.

Risk assessments for inspection and testing of electrical installations

Initial verification requires inspection and testing to be carried out correctly, and in a safe and competent manner.

The table gives the five stages of risk assessment. Use a copy of the table in the activity.

Example of risk assessment during initial verification (partially completed)

1 What are the hazards?	2 Who might be affected? Why? How?	3 a) What are you already doing to reduce the risk of danger and injury?	3 b) What further action is required?	4 How will the risk assessment be implemented?	5 When will the risk assessment be reviewed?
Working on or near live conductors when carrying out live tests in accordance with BS 7671.	Inspectors and testers of the electrical systems and ...	Providing and maintaining correct and appropriate equipment, and ...	Training on equipment likely to be encountered during the tasks (to ensure competence) and ...	Practical assessment for various procedures, such as ...	If new systems are encountered, new test equipment is supplied, new hazards are identified or ...
Isolation to ensure safe inspection and testing of electrical installations during initial verification.					

Reporting of unsafe situations

In order to prevent accidents, you should always report unsafe situations. Causes of unsafe situations could be:

- unsuitable or inappropriate PPE
- inadequate test equipment
- lack of training
- poor awareness of dangers
- carelessness
- incompetence.

Safe use of tools, test equipment and personal protective equipment

You have already covered the need for PPE and the use of correct tools. Now you are going to look at a brief list of equipment and to consider the correct use and maintenance of this equipment during the initial verification process.

The tools and equipment listed below are typical of what may be needed during isolation, inspection and testing:

- locks and keys appropriate to the systems to be worked on (permits to work may be required)
- warning signs, notices and labelling
- approved voltage indicator – see GS38 information listed on page 12
- correct proving unit for the specific voltage indicator
- safety barriers to prevent access by other trades, clients and members of the public
- approved test equipment designed to perform tasks in accordance with BS 7671 – see GS38 information listed on page 12
- PPE appropriate to the work to be undertaken
- hand tools, suitably insulated and maintained.

The GS38 information listed below relates to the design and maintenance of approved electrical test equipment for use by electricians.

- Probes should have:
  - finger barriers
  - an exposed metal tip not exceeding 4 mm; however, it is strongly recommended that this is reduced to 2 mm or less
  - fuse, or fuses, with a low current rating (usually not exceeding 500 mA), or a current-limiting resistor and a fuse.
- Leads should be:
  - adequately insulated
  - colour coded
  - flexible and of sufficient capacity
  - protected against mechanical damage
  - long enough
  - sealed into the body of the voltage detector and should not have accessible exposed conductors, other than the probe tips.

Individual types of test equipment specifications are addressed on pages 47–50.

KEY POINT

You can find GS38 as a 10-page free download on the Health and Safety Executive website: www.hse.gov.uk

Understand the principles and regulatory requirements for inspecting, testing and commissioning electrical systems, equipment and components

WHY YOU NEED TO INSPECT AND TEST ELECTRICAL INSTALLATIONS

Inspection and testing of electrical installations is carried out for initial verification and for periodic inspection and testing:

- *Initial verification* is carried out to ensure that a new circuit (or installation) has been designed, installed, inspected and tested in accordance with BS 7671. This confirms that the installation is in a safe and suitable condition for use.

- *Periodic inspecting and testing* is carried out to assess the on-going safety of the existing electrical installation. This periodic report is *not* a certificate, since a certificate must have all inspection and test criteria in accordance with BS 7671 with no faults or unsatisfactory comments. The periodic report allows for limitations and is used to give an overview of the on-going condition of the installation and whether it is suitable for continued use.

STATUTORY AND NON-STATUTORY DOCUMENTATION, AND REQUIREMENTS

Listed below are some of the documents (with explanations) that affect the processes of initial verification and periodic inspection and testing.

Statutory documents

There are five statutory documents that you need to abide by:

1   The Health and Safety at Work etc Act 1974 (HSW Act). This imposes general duties on employers to ensure the health, safety and welfare at work of all employees. These duties apply to virtually everything in the workplace, including electrical systems and installations.

2   The Electricity at Work Regulations 1989 (EAWR). These are concerned specifically with electrical safety.

3   The Management of Health and Safety at Work Regulations 1999. Regulation 3 requires every employer to carry out an assessment of the risks to workers and any others who may be affected.

ACTIVITY

Guidance Note 3 gives recommended periods of time between tests for different installations. How often should domestic dwellings be tested?

ASSESSMENT GUIDANCE

You will be required to quote both statutory and non-statutory documents that relate to the process of inspection, testing and commissioning. This section will give you the information required to answer written assessments.

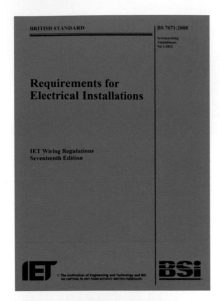

4 The Provision and Use of Work Equipment Regulations 1998. These apply to any machinery, appliance, apparatus or tool used for carrying out any work activity, whether or not electricity is involved. Regulation 5 requires such equipment to be maintained in an efficient state, in working order and in good repair. Also, any maintenance logs for machinery must be kept up to date.

5 Electricity Safety, Quality and Continuity Regulations 2002. These apply to public and consumer safety with regard to electrical distribution and supply authorities.

Non-statutory documents

The following non-statutory documents are important and are referred to throughout this book:

1 The Institution of Engineering and Technology (IET) Wiring Regulations (BS 7671) are based on internationally agreed documents – International Electrotechnical Commission (IEC) harmonised documents – and are international safety rules for electrical installations. A British Standards Institute (BSI) IET-led committee, known as JPEL 64, compiles a UK edition of these rules, selecting regulations specifically for the UK in agreement with a European committee known as CENELEC.

2 The IET On-Site Guide (OSG) to BS 7671 is a simple guide to the requirements from the practical approach of designing, installing, inspecting and testing electrical installations. It can be used as a quick reference guide, but electrical installers should always consult BS 7671 to satisfy themselves of compliance. It is expected that people carrying out work in accordance with this guide are competent to do so.

3 The IET Guidance Note 3: Inspection and Testing to BS 7671 is one of a number of publications prepared by the IET, giving guidance to BS 7671. GN3 is a descriptive guide to the requirements of BS 7671 and provides specific guidance on inspection and testing. Electrical installers should always consult BS 7671 to satisfy themselves of compliance. It is expected that people carrying out work in accordance with this guide are competent to do so.

4 The Guidance Note GS38 – Electrical test equipment for use by electricians (published by the Health and Safety Executive, HSE) was written as a guideline to good practice when using test equipment. It is intended to be followed, in order to reduce the risk of danger and injury when performing electrical tests.

Requirements for initial verification and periodic inspection and testing

Within the non-statutory documentation, there are forms that must be completed for both initial verification and periodic inspections.

Initial verification forms

Initial verification forms must be completed for all new installations and new circuits.

The forms shown here are completed samples. The information given in the forms is not too relevant at this stage, but you do need to know *which* forms are required for satisfactory initial verification:

1 Electrical Installation Certificate

2 Schedule(s) of Inspections

3 Generic Schedule(s) of Test Results

All documents must be completed and authenticated by a competent person(s) before certification is handed over.

The sample documents are based on a new electrical installation in a small office. The office has its own supply and metering system.

Always check you are using the latest forms, as found on the IET website: http://electrical.theiet.org

ACTIVITY

List who is responsible for signing an Electrical Installation Certificate.

ASSESSMENT GUIDANCE

You will be expected to correctly fill in all the forms for reporting and certification, as required.

KEY POINT

Initial Verification can also be recorded using a Minor Electrical Installation Works Certificate but this is restricted to additions or alterations to an existing circuit only. See page 11 for more information.

Form 2   Form No: *SSSS13*....../2

ELECTRICAL INSTALLATION CERTIFICATE
(REQUIREMENTS FOR ELECTRICAL INSTALLATIONS - BS 7671 [IET WIRING REGULATIONS])

DETAILS OF THE CLIENT   *Mr D Roberts*
.... *23 Acacia Avenue* .... Post Code: *SL0 0LT* ....
.... *Sometown, Berks* ....

INSTALLATION ADDRESS   *Unit 3 The Quadrant* ....
.... *Sometown Business Park* ....
.... *Sometown, Berks* .... Post Code: ..... *SL1 0ZZ* ....

DESCRIPTION AND EXTENT OF THE INSTALLATION Tick boxes as appropriate

Description of installation: *Commercial office*

Extent of installation covered by this Certificate: *Full new installation*

(Use continuation sheet if necessary)   see continuation sheet No: ....

New installation	☑
Addition to an existing installation	☐
Alteration to an existing installation	☐

FOR DESIGN
I/We being the person(s) responsible for the design of the electrical installation (as indicated by my/our signatures below), particulars of which are described above, having exercised reasonable skill and care when carrying out the design hereby CERTIFY that the design work for which I/we have been responsible is to the best of my/our knowledge and belief in accordance with BS 7671:2008, amended to .. *2011* .... (date) except for the departures, if any, detailed as follows:

Details of departures from BS 7671 (Regulations 120.3 and 133.5):
None   N/A

The extent of liability of the signatory or the signatories is limited to the work described above as the subject of this Certificate.

For the DESIGN of the installation: **(Where there is mutual responsibility for the design)

Signature: .... *D.Jones* ....Date: *15/08/2013*   Name (IN BLOCK LETTERS): .... *D.JONES* .... Designer No 1

Signature: .... *N/A* ....Date: ....   Name (IN BLOCK LETTERS): .... *N/A* .... Designer No 2**

FOR CONSTRUCTION
I/We being the person(s) responsible for the construction of the electrical installation (as indicated by my/our signatures below), particulars of which are described above, having exercised reasonable skill and care when carrying out the construction hereby CERTIFY that the construction work for which I/we have been responsible is to the best of my/our knowledge and belief in accordance with BS 7671:2008, amended to ..... *2011* ....(date) except for the departures, if any, detailed as follows:

Details of departures from BS 7671 (Regulations 120.3 and 133.5):
None   N/A

The extent of liability of the signatory is limited to the work described above as the subject of this Certificate.

For CONSTRUCTION of the installation:

Signature: .... *T.Smith* ....Date: ... *15/08/2013* ....   Name (IN BLOCK LETTERS): .... *T.SMITH* ....

FOR INSPECTION & TESTING
I/We being the person(s) responsible for the inspection & testing of the electrical installation (as indicated by my/our signatures below), particulars of which are described above, having exercised reasonable skill and care when carrying out the inspection & testing hereby CERTIFY that the work for which I/we have been responsible is to the best of my/our knowledge and belief in accordance with BS 7671:2008, amended to ...... *2011* ....(date) except for the departures, if any, detailed as follows:

Details of departures from BS 7671 (Regulations 120.3 and 133.5):
None   N/A

The extent of liability of the signatory is limited to the work described above as the subject of this Certificate.

For INSPECTION AND TESTING of the installation:

Signature: ...... *G.Wilson* ....Date: .... *15/08/2013* ....   Name (IN BLOCK LETTERS): .... *G.WILSON* ....

NEXT INSPECTION
I/We the designer(s), recommend that this installation is further inspected and tested after an interval of not more than ...... *5* .... years/~~months~~.

Page *1* of ... *4*

Electrical Installation Certificate sample 1 (always check you are using the latest forms, as found on the IET website: http://electrical.theiet.org)

Form 2   Form No: ...*SSSS13*............../2

PARTICULARS OF SIGNATORIES TO THE ELECTRICAL INSTALLATION CERTIFICATE

Designer (No 1)

Name: ....*D.Jones*..   Company: ....*The Electrical Design Partnership*...................

Address: ....*23 High Street*..............................

...............*Sometown, Berks*..........................   Postcode: ...*SL10 0YY*...........Tel No: .*01000 999999*..............

Designer (No 2)
(if applicable)

Name: ....   Company: ....

Address: ....

....   Postcode: ....Tel No: ....

Constructor

Name: .....*T Smith*......................................   Company: ....*T Smith Electrical Installations*.................

Address: ...*Unit 8a Sometown Ind Estate*.............

...............*Sometown, Berks*..........................   Postcode: ......*SL3 0XX*........Tel No: ......*01000 888888*......

Inspector

Name: .....*G Wilson*....................................   Company: .*Wilson and Sons*..

Address: ....*11 Crabtree Row*.........................

...............*Sometown, Berks*..........................   Postcode: .....*SL2 0WW*.......Tel No: ....*01000 777777*.............

SUPPLY CHARACTERISTICS AND EARTHING ARRANGEMENTS Tick boxes and enter details, as appropriate

Earthing arrangements	Number and Type of Live Conductors	Nature of Supply Parameters	Supply Protective Device Characteristics
TN-C ☐ TN-S ☐ TN-C-S ☑ TT ☐ IT ☐ Other sources ☐ of supply (to be detailed on attached schedules)	a.c. ☑ d.c. ☐ 1-phase, 2-wire ☑ 2-wire ☐ 1-phase, 3-wire ☐ 3-wire ☐ 2-phase, 3-wire ☐ other ☐ 3-phase, 3-wire ☐ 3-phase, 4-wire ☐ Confirmation of supply polarity ☑	Nominal voltage, $U/U_0{}^{(1)}$ ....*230*......... V Nominal frequency, $f^{(1)}$ ......*50*...... Hz Prospective fault current, $I_{pf}{}^{(2)}$.*1.41*.. kA External loop impedance, $Z_e{}^{(2)}$ ......*0.34*Ω *(Note: (1) by enquiry, (2) by enquiry or by measurement)*	Type: ...*BS 88-3*..... Rated current..*80*. A

PARTICULARS OF INSTALLATION REFERRED TO IN THE CERTIFICATE Tick boxes and enter details, as appropriate

Means of Earthing

Distributor's facility ☑

Installation earth electrode ☐

Maximum Demand

Maximum demand (load) ....*68*........   kVA / Amps ~Delete as appropriate~

Details of Installation Earth Electrode (*where applicable*)

Type (e.g. rod(s), tape etc)	Location	Electrode resistance to Earth
...........*N/A*............	...........*N/A*....................	...........*N/A*.................... Ω

Main Protective Conductors

Earthing conductor: material ...*Copper*........... csa ....*16*.........mm^2 Continuity and connection verified ☑

Main protective bonding conductors   material ....*Copper*........... csa ....*16*........mm^2 Continuity and connection verified ☑

To incoming water and/or gas service ☑   To other elements: ....*N/A*..........................

Main Switch or Circuit-breaker

BS, Type and No. of poles ......*BS EN 60497-3 (2-pole)*..............   Current rating ....*100*.....A   Voltage rating .....*400*........V

Location ..*Office Suite Consumer Unit*......................................   Fuse rating or setting......*N/A*............ A

Rated residual operating current $I_{\Delta n}$ = ...*N/A*... mA, and operating time of ....*N/A*... ms (at $I_{\Delta n}$) ~(applicable only where an RCD is suitable and is used as a main circuit breaker)~

COMMENTS ON EXISTING INSTALLATION (in the case of an addition or alteration see Section 633):

..........*N/A*..

....

....

....

SCHEDULES

The attached Schedules are part of this document and this Certificate is valid only when they are attached to it.

......*1*...... Schedules of Inspections and ......*1*...... Schedules of Test Results are attached.
~(Enter quantities of schedules attached)~

Page .2.of .4.

Electrical Installation Certificate sample 2 (always check you are using the latest forms, as found on the IET website: http://electrical.theiet.org)

Form 3   Form No: ...SSSS13...../3

SCHEDULE OF INSPECTIONS (for new installation work only)

Methods of protection against electric shock

Both basic and fault protection:

✓	(i)	SELV (note 1)
N/A	(ii)	PELV
N/A	(iii)	Double insulation
N/A	(iv)	Reinforced insulation

Basic protection: (note 2)

✓	(i)	Insulation of live parts
✓	(ii)	Barriers or enclosures
N/A	(iii)	Obstacles (note 3)
N/A	(iv)	Placing out of reach (note 4)

Fault protection:

(i) Automatic disconnection of supply:

✓	Presence of earthing conductor
✓	Presence of circuit protective conductors
✓	Presence of protective bonding conductors
✓	Presence of supplementary bonding conductors
N/A	Presence of earthing arrangements for combined protective and functional purposes
N/A	Presence of adequate arrangements for other sources, where applicable
N/A	FELV
✓	Choice and setting of protective and monitoring devices (for fault and/or overcurrent protection)

(ii) Non-conducting location: (note 5)

N/A	Absence of protective conductors

(iii) Earth-free local equipotential bonding: (note 6)

N/A	Presence of earth-free local equipotential bonding

(iv) Electrical separation: (note 7)

N/A	Provided for **one item** of current-using equipment
N/A	Provided for **more than one item** of current-using equipment

Additional protection:

✓	Presence of residual current devices(s)
✓	Presence of supplementary bonding conductors

Prevention of mutual detrimental influence

✓	(a)	Proximity to non-electrical services and other influences
✓	(b)	Segregation of Band I and Band II circuits or use of Band II insulation
N/A	(c)	Segregation of safety circuits

Identification

✓	(a)	Presence of diagrams, instructions, circuit charts and similar information
✓	(b)	Presence of danger notices and other warning notices
✓	(c)	Labelling of protective devices, switches and terminals
✓	(d)	Identification of conductors

Cables and conductors

✓	Selection of conductors for current-carrying capacity and voltage drop
✓	Erection methods
✓	Routing of cables in prescribed zones
N/A	Cables incorporating earthed armour or sheath, or run within an earthed wiring system, or otherwise adequately protected against nails, screws and the like
N/A	Additional protection provided by 30 mA RCD for cables concealed in walls (where required in premises not under the supervision of a skilled or instructed person)
✓	Connection of conductors
✓	Presence of fire barriers, suitable seals and protection against thermal effects

General

✓	Presence and correct location of appropriate devices for isolation and switching
✓	Adequacy of access to switchgear and other equipment
N/A	Particular protective measures for special installations and locations
✓	Connection of single-pole devices for protection or switching in line conductors only
✓	Correct connection of accessories and equipment
N/A	Presence of undervoltage protective devices
✓	Selection of equipment and protective measures appropriate to external influences
✓	Selection of appropriate functional switching devices

Inspected by ....G. Wilson.......................................   Date .....15/08/2013...

NOTES:
✓   to indicate an inspection has been carried out and the result is satisfactory
N/A   to indicate that the inspection is not applicable to a particular item
An entry must be made in every box.

1. SELV An extra-low voltage system which is electrically separated from Earth and from other systems. The particular requirements of the Regulations must be checked (see Section 414)
2. Method of basic protection - will include measurement of distances where appropriate
3. Obstacles - only adopted in special circumstances (see Regulations 416.2 and 417.2)
4. Placing out of reach - only adopted in special circumstances (see Regulation 417.3)
5. Non-conducting locations - not applicable in domestic premises and requiring special precautions (see Regulation 418.1)
6. Earth-free local equipotential bonding - not applicable in domestic premises, only used in special circumstances (see Regulation 418.2)
7. Electrical separation (see Section 413 and Regulation 418.3)

Schedule of Inspections sample (always check you are using the latest forms, as found on the IET website: http://electrical.theiet.org)

ASSESSMENT GUIDANCE

Remember that you must complete all the boxes on the Schedule of Inspections. Blank boxes or an 'X' are not allowed.

Form 4

Form No: ......1235........../4

GENERIC SCHEDULE OF TEST RESULTS

DB reference no ....Commercial office....
Location ....Unit 3, The Quadrant, SL1 022....
Zs at DB (Ω) ....0.34....
Ipf at DB (kA) ....1.41....
Correct supply polarity confirmed ☑
Phase sequence confirmed (where appropriate) N/A

Details of circuits and/or installed equipment vulnerable to damage when testing ....Downlighter spots – electronic SELV transformer....

Details of test instruments used (state serial and/or asset numbers)
Continuity....Megger multi-function. 10563....
Insulation resistance ...."....
Earth fault loop impedance ...."....
RCD...."....
Earth electrode resistance ....N/A....

Tested by:
Name (Capitals) ....G WILSON....
Signature ....G. Wilson....   Date ....15/08/2013....

Page 4 ... of ...4

Circuit number	Circuit Description	Overcurrent device				Conductor details			Ring final circuit continuity (Ω)			Continuity (Ω) (R₁+R₂) or R₂		Insulation Resistance (MΩ)		Polarity	Test results Zs (Ω)	RCD (ms)		Test button / functionality	Remarks (continue on a separate sheet if necessary)
1	2	BS (EN) 3	type 4	rating (A) 5	breaking capacity (kA) 6	Reference Method 7	Live (mm²) 8	cpc (mm²) 9	r₁ (line) 10	rₙ (neutral) 11	r₂ (cpc) 12	(R₁ + R₂)* 13	R₂ 14	Live-Live 15	Live-E 16	17	Zs 18	@ Δn 19	@ 5 Δn 20	21	22
1	Socket outlets – data	60898	B	20	6	B	2.5	1.5	N/A	N/A	N/A	0.38	N/A	>200	>200	✓	0.75	N/A	N/A	N/A	✓ Checked for compliance
2	Socket outlets – wall	61009	B	32	6	B	2x2.5	2x1.5	0.62	0.62	1.02	0.41	N/A	>200	>200	✓	0.71	85	16	✓	✓ – " –
3	Down lighter spots	60898	B	6	6	B	1.5	1.0	N/A	N/A	N/A	0.56	N/A	>200	>200	✓	0.83	N/A	N/A	N/A	✓ – " –
4	General lighting	60898	B	10	6	B	1.5	1.0	N/A	N/A	N/A	0.48	N/A	>200	>200	✓	0.81	N/A	N/A	N/A	✓ – " –
5	Water heater	60898	B	16	6	B	2.5	1.5	N/A	N/A	N/A	0.11	N/A	>200	>200	✓	0.45	N/A	N/A	N/A	✓ – " –

* Where there are no spurs connected to a ring final circuit this value is also the (R₁ + R₂) of the circuit.

Generic Schedule of Test Results sample (always check you are using the latest forms, as found on the IET website: http://electrical.theiet.org)

Periodic inspection and testing forms

Periodic inspection and testing is carried out to assess the on-going safety of existing electrical installations.

The following forms are blank samples. These forms are *not* classed as certification; they are reports relating to the condition of the electrical installation. The **extent** and **limitations** of the report must be agreed with the person ordering the work. The reasons for these limitations must be validated.

All documents must be completed and authenticated by a competent person(s) before a full report is handed over.

Documentation to be completed for periodic inspection and testing can be found in Appendix 6 of BS 7671:

1   Electrical Installation Condition Report

2   Condition Report Inspection Schedule(s)

3   Generic Schedule(s) of Test Results

Extent

The amount of inspection and testing. For example, in a third floor flat with a single distribution board and eight circuits, the inspection will be visual only without removing covers and testing, and will involve sample tests at the final point of each circuit.

Limitation

A part of the inspection and test process that cannot be done for operational reasons. For example, the main protective bonding connection to the water system, located in the basement, could not be inspected as a key for the room was not available.

ACTIVITY

Using statutory documents as a guide, decide who would be a competent person to carry out testing.

ELECTRICAL INSTALLATION CONDITION REPORT

SECTION A. DETAILS OF THE CLIENT / PERSON ORDERING THE REPORT

Name ....

Address ....

SECTION B. REASON FOR PRODUCING THIS REPORT ....

Date(s) on which inspection and testing was carried out ....

SECTION C. DETAILS OF THE INSTALLATION WHICH IS THE SUBJECT OF THIS REPORT

Occupier ....

Address ....

Description of premises (tick as appropriate)

Domestic ☐ Commercial ☐ Industrial ☐ Other (include brief description) ☐

Estimated age of wiring system ....years

Evidence of additions / alterations Yes ☐ No ☐ Not apparent ☐ If yes, estimate age ....years

Installation records available? (Regulation 621.1) Yes ☐ No ☐   Date of last inspection ....(date)

SECTION D. EXTENT AND LIMITATIONS OF INSPECTION AND TESTING

Extent of the electrical installation covered by this report ....

Agreed limitations including the reasons (see Regulation 634.2) ....

Agreed with: ....

Operational limitations including the reasons (see page no..............) ....

The inspection and testing detailed in this report and accompanying schedules have been carried out in accordance with BS 7671: 2008 (IET Wiring Regulations) as amended to ....

It should be noted that cables concealed within trunking and conduits, under floors, in roof spaces, and generally within the fabric of the building or underground, have **not** been inspected unless specifically agreed between the client and inspector prior to the inspection.

SECTION E. SUMMARY OF THE CONDITION OF THE INSTALLATION

General condition of the installation (in terms of electrical safety) ....

Overall assessment of the installation in terms of its suitability for continued use

SATISFACTORY / UNSATISFACTORY* (Delete as appropriate)

*An unsatisfactory assessment indicates that dangerous (code C1) and/or potentially dangerous (code C2) conditions have been identified.

SECTION F. RECOMMENDATIONS

Where the overall assessment of the suitability of the installation for continued use above is stated as UNSATISFACTORY, I / we recommend that any observations classified as *'Danger present'* (code C1) or *'Potentially dangerous'* (code C2) are acted upon as a matter of urgency. Investigation without delay is recommended for observations identified as *'further investigation required'*.

Observations classified as *'Improvement recommended'* (code C3) should be given due consideration.

Subject to the necessary remedial action being taken, I / we recommend that the installation is further inspected and tested by .... (date)

SECTION G. DECLARATION

I/We, being the person(s) responsible for the inspection and testing of the electrical installation (as indicated by my/our signatures below), particulars of which are described above, having exercised reasonable skill and care when carrying out the inspection and testing, hereby declare that the information in this report, including the observations and the attached schedules, provides an accurate assessment of the condition of the electrical installation taking into account the stated extent and limitations in section D of this report.

Inspected and tested by:

Name (Capitals) ....

Signature ....

For/on behalf of ....

Position ....

Address ....

Date ....

Report authorised for issue by:

Name (Capitals) ....

Signature ....

For/on behalf of ....

Position ....

Address ....

Date ....

SECTION H. SCHEDULE(S)

................schedule(s) of inspection and ....schedule(s) of test results are attached.

The attached schedule(s) are part of this document and this report is valid only when they are attached to it.

SECTION I. SUPPLY CHARACTERISTICS AND EARTHING ARRANGEMENTS

Earthing arrangements	Number and Type of Live Conductors	Nature of Supply Parameters	Supply Protective Device
TN-C ☐	a.c. ☐ d.c. ☐	Nominal voltage, U / $U_0^{(1)}$ V	BS (EN) ....
TN-S ☐	1-phase, 2-wire ☐ 2-wire ☐	Nominal frequency, $f^{(1)}$ ....Hz	Type ....
TN-C-S ☐	2 phase, 3-wire ☐ 3-wire ☐	Prospective fault current, $I_{pf}^{(2)}$ ....kA	Rated current ....A
TT ☐	3 phase, 3-wire ☐	External loop impedance, $Ze^{(2)}$ ....Ω	
IT ☐	3 phase, 4-wire ☐	Note: (1) by enquiry	
	Confirmation of supply polarity ☐	(2) by enquiry or by measurement	

Other sources of supply (as detailed on attached schedule) ☐

SECTION J. PARTICULARS OF INSTALLATION REFERRED TO IN THE REPORT

Means of Earthing   **Details of Installation Earth Electrode** *(where applicable)*

Distributor's facility ☐   Type ....

Installation earth electrode ☐   Location ....

Resistance to Earth ....Ω

Main Protective Conductors

Earthing conductor	Material	Csa ....mm²	Connection / continuity verified ☐
Main protective bonding conductors	Material	Csa ....mm²	Connection / continuity verified ☐
To incoming water service ☐	To incoming gas service ☐	To incoming oil service ☐	To structural steel ☐
To lightning protection ☐	To other incoming service(s) ☐	Specify	

Main Switch / Switch-Fuse / Circuit-Breaker / RCD

	If RCD main switch
Location	Current rating ....A	Rated residual operating current ($I_{Δn}$) ....mA
BS(EN)	Fuse / device rating or setting ....A	Rated time delay ....ms
No of poles	Voltage rating V	Measured operating time(at $I_{Δn}$) ....ms

SECTION K. OBSERVATIONS

Referring to the attached schedules of inspection and test results, and subject to the limitations specified at the *Extent and limitations of inspection and testing* section

No remedial action is required ☐   The following observations are made ☐ (see below):

OBSERVATION(S)	CLASSIFICATION CODE	FURTHER INVESTIGATION REQUIRED (YES / NO)
....	....	....
....	....	....
....	....	....
....	....	....
....	....	....
....	....	....
....	....	....

One of the following codes, as appropriate, has been allocated to each of the observations made above to indicate to the person(s) responsible for the installation the degree of urgency for remedial action.

C1 – Danger present. Risk of injury. Immediate remedial action required

C2 – Potentially dangerous - urgent remedial action required

C3 – Improvement recommended

Electrical Installation Condition Report (always check you are using the latest forms, as found on the IET website: http://electrical.theiet.org)

CONDITION REPORT INSPECTION SCHEDULE FOR DOMESTIC AND SIMILAR PREMISES WITH UP TO 100 A SUPPLY

Note: This form is suitable for many types of smaller installation not exclusively domestic.

OUTCOMES	Acceptable condition ✓	Unacceptable condition State C1 or C2	Improvement recommended State C3	Not verified N/V	Limitation LIM	Not applicable N/A

ITEM NO	DESCRIPTION	OUTCOME (Use codes above. Provide additional comment where appropriate. C1, C2 and C3 coded items to be recorded in Section K of the Condition Report)	Further investigation required? (Y or N)
1.0	**DISTRIBUTOR'S / SUPPLY INTAKE EQUIPMENT**		
1.1	Service cable condition		
1.2	Condition of service head		
1.3	Condition of tails - Distributor		
1.4	Condition of tails - Consumer		
1.5	Condition of metering equipment		
1.6	Condition of isolator (where present)		
2.0	**PRESENCE OF ADEQUATE ARRANGEMENTS FOR OTHER SOURCES SUCH AS MICROGENERATORS (551.6; 551.7)**		
3.0	**EARTHING / BONDING ARRANGEMENTS (411.3; Chap 54)**		
3.1	Presence and condition of distributor's earthing arrangement (542.1.2.1; 542.1.2.2)		
3.2	Presence and condition of earth electrode connection where applicable (542.1.2.3)		
3.3	Provision of earthing / bonding labels at all appropriate locations (514.11)		
3.4	Confirmation of earthing conductor size (542.3; 543.1.1)		
3.5	Accessibility and condition of earthing conductor at MET (543.3.2)		
3.6	Confirmation of main protective bonding conductor sizes (544.1)		
3.7	Condition and accessibility of main protective bonding conductor connections (543.3.2; 544.1.2)		
3.8	Accessibility and condition of all protective bonding connections (543.3.2)		
4.0	**CONSUMER UNIT(S) / DISTRIBUTION BOARD(S)**		
4.1	Adequacy of working space / accessibility to consumer unit / distribution board (132.12; 513.1)		
4.2	Security of fixing (134.1.1)		
4.3	Condition of enclosure(s) in terms of IP rating etc (416.2)		
4.4	Condition of enclosure(s) in terms of fire rating etc (526.5)		
4.5	Enclosure not damaged/deteriorated so as to impair safety (621.2(iii))		
4.6	Presence of main linked switch (as required by 537.1.4)		
4.7	Operation of main switch (functional check) (612.13.2)		
4.8	Manual operation of circuit-breakers and RCDs to prove disconnection (612.13.2)		
4.9	Correct identification of circuit details and protective devices (514.8.1; 514.9.1)		
4.10	Presence of RCD quarterly test notice at or near consumer unit / distribution board (514.12.2)		
4.11	Presence of non-standard (mixed) cable colour warning notice at or near consumer unit / distribution board (514.14)		
4.12	Presence of alternative supply warning notice at or near consumer unit / distribution board (514.15)		
4.13	Presence of other required labelling (please specify) (Section 514)		
4.14	Examination of protective device(s) and base(s); correct type and rating (no signs of unacceptable thermal damage, arcing or overheating) (421.1.3)		
4.15	Single-pole protective devices in line conductor only (132.14.1; 530.3.2)		
4.16	Protection against mechanical damage where cables enter consumer unit / distribution board (522.8.1; 522.8.11)		
4.17	Protection against electromagnetic effects where cables enter consumer unit / distribution board / enclosures (521.5.1)		
4.18	RCD(s) provided for fault protection – includes RCBOs (411.4.9; 411.5.2; 531.2)		
4.19	RCD(s) provided for additional protection - includes RCBOs (411.3.3; 415.1)		

ITEM NO	DESCRIPTION	OUTCOME (Use codes above. Provide additional comment where appropriate. C1, C2 and C3 coded items to be recorded in Section K of the Condition Report)	Further investigation required? (Y or N)
5.0	**FINAL CIRCUITS**		
5.1	Identification of conductors (514.3.1)		
5.2	Cables correctly supported throughout their run (522.8.5)		
5.3	Condition of insulation of live parts (416.1)		
5.4	Non-sheathed cables protected by enclosure in conduit, ducting or trunking (521.10.1) • To include the integrity of conduit and trunking systems (metallic and plastic)		
5.5	Adequacy of cables for current-carrying capacity with regard for the type and nature of installation (Section 523)		
5.6	Coordination between conductors and overload protective devices (433.1; 533.2.1)		
5.7	Adequacy of protective devices: type and rated current for fault protection (411.3)		
5.8	Presence and adequacy of circuit protective conductors (411.3.1.1; 543.1)		
5.9	Wiring system(s) appropriate for the type and nature of the installation and external influences (Section 522)		
5.10	Concealed cables installed in prescribed zones (see Section D. *Extent and limitations*) (522.6.101)		
5.11	Concealed cables incorporating earthed armour or sheath, or run within earthed wiring system, or otherwise protected against mechanical damage from nails, screws and the like (see Section D. *Extent and limitations*) (522.6.101; 522.6.103)		
5.12	Provision of additional protection by RCD not exceeding 30 mA: • for all socket-outlets of rating 20 A or less provided for use by ordinary persons unless an exception is permitted (411.3.3) • for supply to mobile equipment not exceeding 32 A rating for use outdoors (411.3.3) • for cables concealed in walls or partitions (522.6.102; 522.6.103)		
5.13	Provision of fire barriers, sealing arrangements and protection against thermal effects (Section 527)		
5.14	Band II cables segregated / separated from Band I cables (528.1)		
5.15	Cables segregated / separated from communications cabling (528.2)		
5.16	Cables segregated / separated from non-electrical services (528.3)		
5.17	Termination of cables at enclosures – indicate extent of sampling in Section D of the report (Section 526) • Connections soundly made and under no undue strain (526.6) • No basic insulation of a conductor visible outside enclosure (526.98) • Connections of live conductors adequately enclosed (526.5) • Adequately connected at point of entry to enclosure (glands, bushes etc.) (522.8.5)		
5.18	Condition of accessories including socket-outlets, switches and joint boxes (621.2 (iii))		
5.19	Suitability of accessories for external influences (512.2)		
6.0	**LOCATION(S) CONTAINING A BATH OR SHOWER**		
6.1	Additional protection for all low voltage (LV) circuits by RCD not exceeding 30 mA (701.411.3.3)		
6.2	Where used as a protective measure, requirements for SELV or PELV met (701.414.4.5)		
6.3	Shaver sockets comply with BS EN 61558-2-5 formally BS 3535 (701.512.3)		
6.4	Presence of supplementary bonding conductors, unless not required by BS 7671 2008 (701.415.2)		
6.5	Low voltage (e.g. 230 volt) socket-outlets sited at least 3 m from zone 1 (701.512.3)		
6.6	Suitability of equipment for external influences for installed location in terms of IP rating (701.512.2)		
6.7	Suitability of equipment for installation in a particular zone (701.512.3)		
6.8	Suitability of current-using equipment for particular position within the location (7C1.55)		
7.0	**OTHER PART 7 SPECIAL INSTALLATIONS OR LOCATIONS**		
7.1	List all other special installations or locations present, if any. (Record separately the results of particular inspections applied.)		

Inspected by:
Name (Capitals) ....   Signature ....   Date ....

Condition Report Inspection Schedule(s) (always check you are using the latest forms, as found on the IET website: http://electrical.theiet.org)

Form 4

Form No: ..../4

GENERIC SCHEDULE OF TEST RESULTS

DB reference no ....
Location ....
Zs at DB (Ω) ....
Ipf at DB (kA) ....
Correct supply polarity confirmed ☐
Phase sequence confirmed (where appropriate) ☐

Details of circuits and/or installed equipment vulnerable to damage when testing ....

Details of test instruments used (state serial and/or asset numbers)
Continuity ....
Insulation resistance ....
Earth fault loop impedance ....
RCD ....
Earth electrode resistance ....

Tested by:
Name (Capitals) ....
Signature ....   Date ....

Circuit details										Ring final circuit continuity (Ω)			Continuity (Ω) (R₁ + R₂) or R₂		Insulation Resistance (MΩ)			Test results					
	Overcurrent device			Conductor details															Z_s (Ω)	RCD (ms)			Remarks (continue on a separate sheet if necessary)
Circuit Description	Circuit number	BS (EN)	type	rating (A)	breaking capacity (kA)	Reference Method	Live (mm²)	cpc (mm²)	r_1 (line)	r_n (neutral)	r_2 (cpc)	$(R_1 + R_2)$ *	R_2	Live-Live	Live-E	Polarity		@ $I\Delta n$	@ 5$I\Delta n$	Test button operation	
	1	2	3	4	5	6	7	8	9	10	11	12	13	14	15	16	17	18	19	20	21	22

* Where there are no spurs connected to a ring final circuit this value is also the (R₁ + R₂) of the circuit.

Generic Schedule(s) of Test Results (always check you are using the latest forms, as found on the IET website: http://electrical.theiet.org)

1 List:
   a) **three** statutory documents relating to initial verification and periodic testing
   b) **three** non-statutory documents relating to initial verification and periodic testing.

2 State the BS 7671 documents that need to be completed for:
   a) an initial verification
   b) periodic inspection and testing.

3 Which one of the following categories of person can authenticate an Electrical Installation Certificate (as stated on the form)?
   a) ordinary
   b) instructed
   c) skilled
   d) competent

ASSESSMENT GUIDANCE

Remember to double check all forms. All parts must be completed.

Assessment criteria

2.3 Specify the information that is required to correctly conduct the initial verification of an electrical installation in accordance with the IET Wiring Regulations and IET Guidance Note 3.

ASSESSMENT GUIDANCE

Diversity takes into account the fact that not all the loads in a building will be on at the same time. How many lights, out of the maximum, are on in your house at any one time in the evening?

INFORMATION REQUIRED TO CARRY OUT INITIAL VERIFICATION OF AN ELECTRICAL INSTALLATION

Before carrying out an initial verification of an electrical installation, you need the following information, in accordance with the IET Wiring Regulations and IET Guidance Note 3:

1   maximum demand and diversity (Regulation 311)

2   conductor and system earthing arrangements (TN or TT System) (Regulation 312)

3   nominal voltage, current, frequency, prospective short-circuit current and external earth fault loop impedance (Z_e) (Regulation 312)

4   compatibility of characteristics (listed in Regulation 313)

5   diagrams, documents, plans and design criteria for the building (Regulation 514.9.1).

Note that the above regulation numbers are from BS 7671.

Maximum demand and diversity (Regulation 311)

The maximum demand of a new installation must be assessed. Appendix A of the IET OSG provides details of the methods used to calculate the maximum demand and diversity of small installations. The assessed demand after diversity must be inserted on the electrical installation certificate.

Conductor and system earthing arrangements (TN or TT system) (Section 312)

Understanding the system earthing arrangements is one of the most important parts of the inspection and testing process. This is covered in greater detail on pages 75–81.

Nominal voltage, current, frequency, prospective fault current and external earth fault loop impedance (Z_e) (Section 312)

The following values are for low voltage public electricity supply systems:

- the nominal voltage to earth (V_0) is 230 Volts
- the current is alternating current (a.c.)
- the frequency is 50 Hz
- the prospective fault current (PFC) will vary with every installation.
- the external earth fault loop impedance will vary with every system, supply arrangements and, therefore, installation.

Compatibility of characteristics (listed in Chapter 33)

An assessment shall be made of any characteristics of equipment likely to have harmful effects on other electrical equipment or other services, or likely to impair the supply.

A list of design considerations can be found at Section 331 of BS 7671.

The list is very technical, including overvoltages, starting currents, harmonic currents, power factor and many more compatibility design considerations.

Diagrams, documents, plans and design criteria for the building (Regulation 514.9.1)

The inspection and certification process requires documentation to be available for the person carrying out the work.

Regulation 514.9.1 lists the following:

> A legible diagram, chart or table or equivalent form of information shall be provided indicating:
>
> (i) the type and composition of each circuit (points of utilisation served, number and size of conductors, type of wiring)
>
> (ii) the method used for compliance with Regulation 410.3.2, for basic protective measures and independent fault protection. Additional protection may need to be considered under certain conditions
>
> (iii) the information necessary for the identification of each device performing the functions of protection, isolation and switching, and its location
>
> (iv) any circuit or equipment vulnerable to a typical test.
>
> For simple installations the foregoing information may be given in a schedule. A durable copy of the schedule relating to a distribution board shall be provided within or adjacent to each distribution board.

ASSESSMENT GUIDANCE

You may be asked to name the information required before inspection and testing can commence.

ACTIVITY

List five items of information that should be made available to the person who is inspecting and testing.

Below is a typical example of a schedule that can be used to highlight the distribution board circuits and information.

Form 4

Form No: ....1235........./4

GENERIC SCHEDULE OF TEST RESULTS

DB reference no ......*Commercial office*............
Location ..*Unit 3, The Quadrant, SL1 0ZZ*..............
Zs at DB (Ω) ....*0.34*.............
I_{pf} at DB (kA) ....*1.41*............
Correct supply polarity confirmed ☑
Phase sequence confirmed (where appropriate) [N/A]

Details of circuits and/or installed equipment vulnerable to damage when testing ....*Downlighters spots – electronic SELV transformers*...............

Details of test instruments used (state serial and/or asset numbers)
Continuity.....................*Megger multi-function 10S63*............
Insulation resistance ...." ....
Earth fault loop impedance ...." ....
RCD..........................." ....
Earth electrode resistance ....*N/A*...........

Tested by:
Name (Capitals) ....*G WILSON*...............
Signature ....*G. Wilson*............... Date ......*15/08/2013*.........

Circuit number	Circuit Description	BS (EN)	type	rating (A)	breaking capacity (kA)	Reference Method	Live (mm²)	cpc (mm²)	r₁ (line)	rₙ (neutral)	r₂ (cpc)	(R₁ + R₂) *	R₂	Live-Live	Live-E	Polarity	Z_s (Ω)	@ ∆n	@ 5 ∆n	Test button / functionality	Remarks (continue on a separate sheet if necessary)
																			(ms)		
1	2	3	4	5	6	7	8	9	10	11	12	13	14	15	16	17	18	19	20	21	22
1	Socket outlets - dado	60898	B	20	6	B	2.5	1.5	N/A	N/A	N/A	0.38	N/A	>200	>200	✓	0.75	N/A	N/A	N/A	✓ Checked for compliance
2	Socket outlets - wall	61009	B	32	6	B	2×2.5	2×1.5	0.62	0.62	1.02	0.41	N/A	>200	>200	✓	0.71	85	16	✓	✓ - " -
3	Down lighter spots	60898	B	6	6	B	1.5	1.0	N/A	N/A	N/A	0.56	N/A	>200	>200	✓	0.83	N/A	N/A	N/A	✓ - " -
4	General lighting	60898	B	10	6	B	1.5	1.0	N/A	N/A	N/A	0.48	N/A	>200	>200	✓	0.81	N/A	N/A	N/A	✓ - " -
5	Water heater	60898	B	16	6	B	2.5	1.5	N/A	N/A	N/A	0.11	N/A	>200	>200	✓	0.45	N/A	N/A	N/A	✓ - " -

Ring final circuit continuity (Ω) — Continuity (Ω) (R₁ + R₂) or R₂ — Insulation Resistance (MΩ) — RCD

Circuit details — Overcurrent device — Conductor details

Test results

* Where there are no spurs connected to a ring final circuit this value is also the (R₁ + R₂) of the circuit.

Page 4 ... of ...4

Generic Schedule of Test Results (always check you are using the latest forms, as found on the IET website: http://electrical.theiet.org)

Understand the regulatory requirements and procedures for completing the inspection of electrical installations

HUMAN SENSES AND INSPECTION

Human senses are vital to the inspection process. You need to know which senses are involved and be able to explain how to use them.

Assessment criteria

3.2 State how human senses (sight, touch etc) can be used during the inspection process

Using senses during inspection of electrical installations

Inspection is normally done with that part of the installation under inspection disconnected from the supply, in accordance with Regulation 611.1. The examples below show how to apply the human senses during the stages of initial verification. Those stages highlighted in red usually occur when the installation is switched on.

Sight

Sight is the most extensively used sense. It is used when:

1 connecting conductors

2 identifying circuits

3 routing cables

4 labelling circuits

5 correctly connecting accessories

6 burning occurs, due to bad connection.

Touch

Touch is used in:

1 connection of conductors (using a screwdriver to check the connections)

2 correct connection to equipment (for example, in the physical connection to a water pipe)

3 erection methods (for example, fixings for distribution boards (DBs)/conduits/trunking)

4 checking of barriers and enclosures (check for IPXXB and IPXXD in accordance with Regulation 416.2)

ASSESSMENT GUIDANCE

Touch is sometimes listed as a means of detecting hot cables. It would be very difficult for fingers or the hand to detect the temperature of a cable accurately.

KEY POINT

There are five senses. Taste is not used in the inspection and testing process, but hearing, touch, sight and smell are used.

5 checking for careless work methods (sharp edges in conduit)

6 detecting overheating of equipment.

Sound

Listening can detect:

1 arcing caused by loading, for example when switching on and off fluorescent fittings

2 equipment noise, such as a motor bearing problem that causes a loud grinding sound

3 arcing caused by insecure connections, for example at an accessory, junction box or distribution board.

Smell

Smell is used to detect:

1 equipment that is overheating

2 loose connections under load conditions

3 burning of adjacent building materials (for example, recessed lights without fire protection).

ASSESSMENT GUIDANCE

During your assessments, you may be asked which human senses you use during the process of inspection and testing.

Assessment criteria

3.3 State the items of an electrical installation that should be inspected in accordance with IET Guidance Note 3

ACTIVITY

Using the symbols in BS 7671, show the relationship between the design current, the protective device rating and the cable rating (current-carrying capacity).

INSPECTION ITEMS EXPLAINED

Some of the important items of an electrical installation that should be inspected are briefly explained below. Other items are explained on pages 123–131.

Items you should inspect

Connection of conductors

Every connection between a conductor and equipment (or another conductor) should provide durable electrical continuity and adequate mechanical strength.

Identification of cables and conductors

Check that each core or bare conductor is identified as necessary.

Routing of cables

Cables and their cable management systems should be designed and installed to take into account the mechanical stresses that users will place on the installation.

Cable selection

Where practicable, the cable size should be assessed against the protective arrangement.

Accessories and equipment

Correct connection (suitability and polarity) and environmental conditions, such as the presence of dust or moisture, must be checked.

Selection and erection to minimise the spread of fire

Fire barriers, suitable seals and/or protection against thermal effects should be provided. These checks should be carried out during the installation process.

Basic protection

Basic protection is most commonly provided by insulation of live parts and/or barriers and enclosures:

- Confirmation of the insulation of live parts should be carried out to ensure no damage has occurred during the installation process.
- Barriers and enclosures need to be checked to ensure a protection level of at least IPXXB or IP2X. For horizontal top surfaces of readily accessible enclosures, you must ensure at least IPXXD or IP4X.

IP codes designated numbering and lettering system

The illustrations and tables that follow explain the IP codes designated numbering and lettering system.

IP code for ingress protection

Where a characteristic numeral does not have to be specified, it can be replaced by the letter 'X' ('XX' if both numbers are omitted).

| | IP | 2 | 3 | C | H |

Code letters
(international protection)

First characteristic numeral
(numerals 0 to 6, or letter X)

Second characteristic numeral
(numerals 0 to 8, or letter X)

Additional letter (optional)
(letters A, B, C, D)

Supplementary letter (optional)
(letters H, M, S, W)

IP codes designated numbering and lettering system

ACTIVITY

How can polarity be checked without the use of a test instrument?

ASSESSMENT GUIDANCE

You will be surprised at what gets into electrical equipment. Make sure all entries are correctly sealed or comply with the relevant IP codes.

IP characteristic numerals

The IP characteristic numerals are shown in the table.

▼ **Table B1** IP code characteristic numerals

First characteristic numeral		Second characteristic numeral	
(a) Protection of persons against access to hazardous parts inside enclosures (b) Protection of equipment against ingress of solid foreign objects		Protection of equipment against ingress of water	
No.	**Degree of protection**	**No.**	**Degree of protection**
0	(a) Not protected (b) Not protected	0	Not protected
1	(a) Protection against access to hazardous parts with the back of the hand (b) Protection against foreign solid objects of 50 mm diameter and greater	1	Protection against vertically falling water drops
2	(a) Protection against access to hazardous parts with a finger (b) Protection against solid foreign objects of 12.5 mm diameter and greater	2	Protected against vertically falling water drops when enclosure tilted up to 15°. Vertically falling water drops shall have no harmful effect when the enclosure is tilted at any angle up to 15° from the vertical
3	(a) Protection against contact by tools, wires or such like more than 2.5 mm thick (b) Protection against solid foreign objects of 2.5 mm diameter and greater	3	Protected against water spraying at an angle up to 60° on either side of the vertical
4	(a) As 3 above but against contact with a wire or strips more than 1.0 mm thick (b) Protection against solid foreign objects of 1.0 mm diameter and greater	4	Protected against water splashing from any direction

IP characteristic numerals (from Guidance Note 1: Selection and Erection of Equipment, IET)

For a code such as IPXXD, you can ignore the numerical values in the table above and simply refer to the additional letter table illustrated here.

▼ **Table B2** IP code additional letters

Additional letter	Brief description of protection
A	Protected against access with the back of the hand (minimum 50 mm diameter sphere) (adequate clearance from live parts)
B	Protected against access with a finger (minimum 12 mm diameter test finger, 80 mm long) (adequate clearance from live parts)
C	Protected against access with a tool (minimum 2.5 mm diameter tool, 100 mm long) (adequate clearance from live parts)
D	Protected against access with a wire (minimum 1 mm diameter wire, 100 mm long) (adequate clearance from live parts)

IP letter codes (from Guidance Note 1: Selection and Erection of Equipment, IET)

ACTIVITY

Which IP code(s) apply to preventing entry by a 1 mm wire or greater?

ACTIVITY

Which IP code relates to the BS Finger: IP4X or IPXXB?

British Standard finger (IPXXB)

Protection against access with a finger can be assessed using the model shown in the picture.

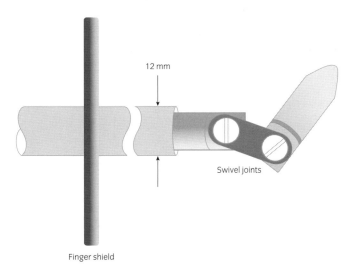

12 mm

Swivel joints

Finger shield

British Standard finger (IPXXB)

Remember that the IP codes are used in conjunction with basic protection, as a guide to protection against contact with live parts and ingress of foreign bodies.

Note also that providing obstacles and placing installations out of reach are methods of protection against contact, but they can only be used as a method of protection in a controlled and supervised environment.

Fault protection

Fault protection inspection involves ensuring that earthing arrangements are in place. Part of this process includes automatic disconnection of supply, selection of protective devices and disconnection times, which will undergo further inspection and verification as you progress through the tests.

Specialised systems

Specialised systems are rarely used. They can be found in Section 418 of BS 7671 and include:

- non-conducting location
- earth-free local equipotential bonding
- electrical separation for the supply to more than one item of current-using equipment.

These systems are normally not applicable (use 'n/a') when filling in the Inspection Schedule.

Electrical separation for the supply to one item of equipment is more common. This may include the shaver point in a bathroom.

Drip-proof IPX2

Rain proof IPX3

Splashproof  IPX4

Jet proof IPX5

Protected against immersion in water IPX7

Codes and signs used on equipment where water is likely to be present

Prevention of mutual detrimental influence

Regulations 132.11 and 515.1 require that the electrical installation and its equipment shall not cause detriment to other electrical and non-electrical installations. Equally, other non-electrical services shall not have a detrimental impact on electrical installations. The inspector is advised to step back and think about other systems when carrying out the inspection – for example, cables under floorboards that touch central heating pipes.

Protective devices, labelling, warning notices and adequate access

These can be observed at the distribution board or consumer unit.

Selection of equipment and protective measures appropriate to external influences

Checks must be made to ensure that equipment has been manufactured to withstand the environmental conditions.

If water is likely to be present, signs and codes such as those shown in the diagram should be present on the equipment.

Erection methods

Chapter 52 of BS 7671 contains detailed requirements on selection and erection. Fixings of switchgear, cables, conduit and fittings must be adequate for the environment and a detailed visual inspection is required during the erection stages, as well as at completion.

INSPECTION PROCESS FOR INITIAL VERIFICATION

All relevant parts of an installation must be inspected during initial verification. In all types of inspection, you must be able to identify which parts of the Schedule of Inspections (see page 36) are relevant and complete the boxes with a tick (✔) or not applicable (n/a).

No boxes on the Schedule of Inspections should be left blank or marked as unsatisfactory (✗).

Assessing the plan of an electrical system to be installed

The location plan of a small industrial unit and the accompanying information shown here and on a larger scale in Appendix 1 should allow you to practise making decisions about completing a Schedule of Inspections. Use this information as you complete the activity on page 34.

Use this information as you complete the activity on page 34.

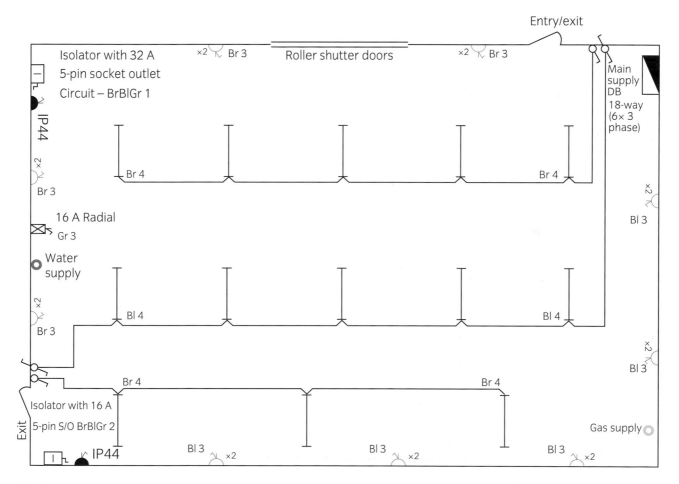

The location plan is a floor plan drawing of small industrial unit. It shows a basic outline of the electrical wiring design. The unit is one of 10 similar units on a small industrial estate.

The supply is a TN-C-S/(**PME**) system. The designer has specified that each unit has a maximum demand allowance of 100 amps per phase.

The wiring is in single-core 70 °C thermoplastic insulated cables, which are installed in a combination of metal and PVC conduit and trunking. There are no **SELV/PELV** circuits to be considered at the time of the inspection.

SELV

Separated extra-low voltage circuit.

PELV

Protective extra-low voltage circuit.

FELV

Functional extra-low voltage circuit (requirements can be found in Section 411.7 of BS 7671).

ACTIVITY

Use the information on pages 33–36, Appendix 1 (page 227) and Appendix 2 (page 228) to complete this activity, which is an exercise in inspection of a small industrial unit.

- You are an electrical inspector, responsible for inspecting the initial verification of the small industrial unit. During the installation process, you have visited the site and inspected the unit at relevant times in accordance with Regulation 610.1 of BS 7671.
- The appropriate design criteria, electrical plans and information that are relevant to the unit have been made available in accordance with Regulations 311, 312, 313 and 514.9.1 (see Regulation 610.2).
- You are a competent person, as required by Regulation 610.5.

Carry out the inspection process of the small industrial unit:

- The Inspection Schedule checklist below is taken from Regulation 611.3 of BS 7671. It has been put in a sequence that matches the Schedule of Inspections, as shown on page 36.
- Use the Inspection Schedule checklist to fill in a copy of the Schedule of Inspections on page 36, for the small industrial unit. The Schedule of Inspections is taken from Appendix 6 of BS 7671.
- When you have completed the Schedule of Inspections, compare your results with the checklist on page 37.

INSPECTION SCHEDULE CHECKLIST

Methods of protection against electric shock

Both basic and fault protection:
(i) SELV (ii) PELV (iii) Double insulation (iv) Reinforced insulation

Basic protection:
(i) Insulation of live parts (ii) Barriers or enclosures (iii) Obstacles (iv) Placing out of reach

Fault protection:

(i) Automatic disconnection of supply:
Presence of earthing conductor
Presence of circuit protective conductors
Presence of protective bonding conductors
Presence of supplementary bonding conductors
Presence of earthing arrangements for combined protective and functional purposes
Presence of adequate arrangements for other sources, where applicable
FELV
Choice and setting of protective and monitoring devices (for fault and/or overcurrent protection)

(ii) Non-conducting location:
Absence of protective conductors

(iii) Earth-free local equipotential bonding:
Presence of earth-free local equipotential bonding

(iv) Electrical separation:
Provided for one item of current-using equipment
Provided for more than one item of current-using equipment

Additional protection:
Presence of residual current device(s)
Presence of supplementary bonding conductors

Prevention of mutual detrimental influence:
(a) Proximity to non-electrical services and other influences
(b) Segregation of Band I and Band II circuits or use of Band II insulation
(c) Segregation of safety circuits

Identification
(a) Presence of diagrams, instructions, circuit charts and similar information
(b) Presence of danger notices and other warning notices
(c) Labelling of protective devices, switches and terminals
(d) Identification of conductors

Cables and conductors
Selection of conductors for current-carrying capacity and voltage drop
Erection methods
Routing of cables in prescribed zones
Cables incorporating earthed armour or sheath, or run within an earthed wiring system, or otherwise adequately protected against nails, screws and the like
Additional protection provided by 30 mA RCD for cables concealed in walls (where required in premises not under the supervision of a skilled or instructed person)
Connection of conductors
Presence of fire barriers, suitable seals and protection against thermal effects

General
Presence and correct location of appropriate devices for isolation and switching
Adequacy of access to switchgear and other equipment
Particular protective measures for special installations and locations
Connection of single-pole devices for protection or switching in line conductors only
Correct connection of accessories and equipment
Presence of undervoltage protective devices
Selection of equipment and protective measures appropriate to external influences
Selection of appropriate functional switching devices

> **ASSESSMENT GUIDANCE**
>
> Circuit charts are essential but often poorly done, especially in domestic installations. Make sure yours are clear and precise.

The Schedule of Inspections

As part of the inspection you will be required to complete a Schedule of Inspections (for new installation work only) as shown on the next page and in Appendix 6 of BS 7671 in relationship to Regulation 610.6.

Always check you are using the latest forms, as found on the IET website: http://electrical.theiet.org

SCHEDULE OF INSPECTIONS (for new installation work only)

Methods of protection against electric shock

Both basic and fault protection:

- ☐ (i) SELV
- ☐ (ii) PELV
- ☐ (iii) Double insulation
- ☐ (iv) Reinforced insulation

Basic protection:

- ☐ (i) Insulation of live parts
- ☐ (ii) Barriers or enclosures
- ☐ (iii) Obstacles
- ☐ (iv) Placing out of reach

Fault protection:

(i) Automatic disconnection of supply:

- ☐ Presence of earthing conductor
- ☐ Presence of circuit protective conductors
- ☐ Presence of protective bonding conductors
- ☐ Presence of supplementary bonding conductors
- ☐ Presence of earthing arrangements for combined protective and functional purposes
- ☐ Presence of adequate arrangements for other sources, where applicable
- ☐ FELV
- ☐ Choice and setting of protective and monitoring devices (for fault and/or overcurrent protection)

(ii) Non-conducting location:

- ☐ Absence of protective conductors

(iii) Earth-free local equipotential bonding:

- ☐ Presence of earth-free local equipotential bonding

(iv) Electrical separation:

- ☐ Provided for **one item** of current-using equipment
- ☐ Provided for **more than one item** of current-using equipment

Additional protection:

- ☐ Presence of residual current device(s)
- ☐ Presence of supplementary bonding conductors

Prevention of mutual detrimental influence

- ☐ (a) Proximity to non-electrical services and other influences
- ☐ (b) Segregation of Band I and Band II circuits or use of Band II insulation
- ☐ (c) Segregation of safety circuits

Identification

- ☐ (a) Presence of diagrams, instructions, circuit charts and similar information
- ☐ (b) Presence of danger notices and other warning notices
- ☐ (c) Labelling of protective devices, switches and terminals
- ☐ (d) Identification of conductors

Cables and conductors

- ☐ Selection of conductors for current-carrying capacity and voltage drop
- ☐ Erection methods
- ☐ Routing of cables in prescribed zones
- ☐ Cables incorporating earthed armour or sheath, or run within an earthed wiring system, or otherwise adequately protected against nails, screws and the like
- ☐ Additional protection provided by 30 mA RCD for cables concealed in walls (where required in premises not under the supervision of a skilled or instructed person)
- ☐ Connection of conductors
- ☐ Presence of fire barriers, suitable seals and protection against thermal effects

General

- ☐ Presence and correct location of appropriate devices for isolation and switching
- ☐ Adequacy of access to switchgear and other equipment
- ☐ Particular protective measures for special installations and locations
- ☐ Connection of single-pole devices for protection or switching in line conductors only
- ☐ Correct connection of accessories and equipment
- ☐ Presence of undervoltage protective devices
- ☐ Selection of equipment and protective measures appropriate to external influences
- ☐ Selection of appropriate functional switching devices

Inspected by ....   Date ....

NOTES:
- ✓ to indicate an inspection has been carried out and the result is satisfactory
- **N/A** to indicate that the inspection is not applicable to a particular item

Checklist of inspection items relevant to the small industrial unit

In the activity on page 34 you were asked to consider a floor plan for a small industrial unit and design criteria listed in an Inspection Schedule checklist, taken from Regulation 611.3. The areas from the Inspection Schedule checklist that need to be inspected and should be completed on the Schedule of Inspections are listed below. Other criteria are not applicable (n/a). Compare your results with those listed here.

Basic protection:
(i) Insulation of live parts (ii) Barriers or enclosures

(i) Automatic disconnection of supply:
Presence of earthing conductor
Presence of circuit protective conductors
Presence of protective bonding conductors
Presence of supplementary bonding conductors
Choice and setting of protective and monitoring devices (for fault and/or
    overcurrent protection)

Additional protection:
Presence of residual current device(s)
Presence of supplementary bonding conductors

Prevention of mutual detrimental influence;
(a) Proximity to non-electrical services and other influences

Cables and conductors
Selection of conductors for current-carrying capacity and voltage drop
Erection methods
Routing of cables in prescribed zones
Cables incorporating earthed armour or sheath, or run within an earthed
    wiring system, or otherwise adequately protected against nails, screws
    and the like
Connection of conductors

General
Presence and correct location of appropriate devices for isolation and
    switching
Adequacy of access to switchgear and other equipment
Connection of single-pole devices for protection or switching in line
    conductors only
Correct connection of accessories and equipment
Selection of equipment and protective measures appropriate to external
    influences
Selection of appropriate functional switching devices

3.4 Specify the requirements for the inspection of the following:

- earthing conductors
- circuit protective conductors
- protective bonding conductors:
  - ☐ main bonding conductors
  - ☐ supplementary bonding conductors
- isolation
- type and rating of overcurrent protective devices

ASSESSMENT GUIDANCE

The electricity isolating switch allows the consumer unit to be totally isolated. It also allows the CU to be changed without contacting the meter owner.

ACTIVITY

Assuming that a 100 A fuse is fitted to the service head, what size consumer's tails would normally be installed in this installation?

SmartScreen Unit 307

Handout 4 and Worksheet 8

INSPECTION REQUIREMENTS

By referring to this illustration, you can assess some of the inspection requirements below.

Remember that this relates to an inspection. All that you need to do is verify the correct connection and size (cross-sectional area) of the conductors. For testing the continuity of these conductors, please see pages 52–65.

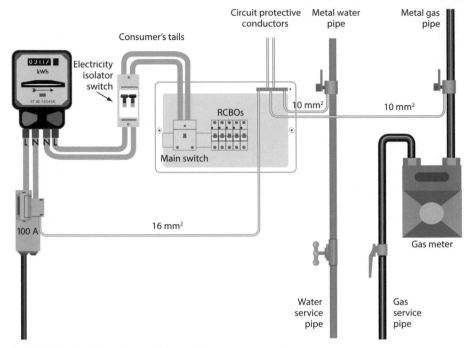

Typical single-phase domestic supply

Visual inspection

Visual checks must be carried out for inspection of the items listed below.

Presence of earthing conductor (shown as a 16 mm² cable)

- Check that the cable is continuous from the origin of the supply to the Main Earthing Terminal (MET), consumer unit or distribution board.
- Ensure that the connections are secure.
- Ensure that the cable fixings are adequate.
- Verify the cross-sectional area of the conductor and compare with calculation or selection in accordance with Regulation 543.1.1.

- Ensure that the supply authority has supplied an earth connection at the origin (for TN-S and TN-C-S systems).
- Ensure that the earth electrode is installed correctly and that the cables are installed in accordance with Table 54.1 of BS 7671 for minimum cross-sectional area (TT system).

Presence of main protective bonding conductors (shown as a 10 mm² cable)

- Check that the cable is continuous from the MET to the water, gas structural metalwork and any other extraneous conductive parts in accordance with Regulation 411.3.1.2. (Where the cable is hidden for part of its length, this must be verified by test.)
- Ensure that the connections are secure.
- Ensure that the cable fixings are adequate.
- Verify the cross-sectional area of the conductor and compare with Table 54.8 of BS 7671 for TN-C-S systems.
- Verify the cross-sectional area of the conductor and compare with Regulation 544.1.1 for TN-S and TT systems.

Presence of supplementary protective bonding conductors

- Ensure the conductors are appropriately installed.
- Ensure that the connections are secure.
- Ensure that the cable fixings are adequate.
- Verify that the conductors are sized in accordance with Regulation section 544.2.3: 'A supplementary bonding conductor connecting two extraneous conductive parts shall have a cross-sectional area not less than 2.5 mm² if sheathed or otherwise provided with mechanical protection or 4 mm² if mechanical protection is not provided.'

Presence of circuit protective conductors (cpc)

- Ensure the conductors are appropriately installed (if required).
- Ensure that the connections are secure.
- Ensure that the cable fixings are adequate.
- Verify that the conductors are selected in accordance with Table 54.7 of BS 7671 or calculated by using Regulation 543.1.3.

$$S = \frac{\sqrt{I^2 t}}{k}$$

ACTIVITY

1 Select the minimum cross-sectional area (csa) of a copper earthing conductor from the table below (Table 54.7 of BS 7671), if the supply cables (tails) for a three-phase +N distribution board are 50 mm² copper conductors (TN-C-S system earth).

TABLE 54.7 –
Minimum cross-sectional area of protective conductor
in relation to the cross-sectional area of associated line conductor

Cross-sectional area of line conductor S	Minimum cross-sectional area of the corresponding protective conductor	
	If the protective conductor is of the same material as the line conductor	If the protective conductor is not of the same material as the line conductor
(mm²)	(mm²)	(mm²)
$S \leq 16$	S	$\frac{k_1}{k_2} \times S$
$16 < S \leq 35$	16	$\frac{k_1}{k_2} \times 16$
$S > 35$	$\frac{S}{2}$	$\frac{k_1}{k_2} \times \frac{S}{2}$

Table 54.7 of BS 7671:2008 (2011)

2 Select the minimum csa of the corresponding main protective bonding conductors for the PME system by using the table below (Table 54.8 of BS 7671).

TABLE 54.8 –
Minimum cross-sectional area of the main protective bonding conductor
in relation to the neutral of the supply

NOTE: Local distributor's network conditions may require a larger conductor.

Copper equivalent cross-sectional area of the supply neutral conductor	Minimum copper equivalent[*] cross-sectional area of the main protective bonding conductor
35 mm² or less	10 mm²
over 35 mm² up to 50 mm²	16 mm²
over 50 mm² up to 95 mm²	25 mm²
over 95 mm² up to 150 mm²	35 mm²
over 150 mm²	50 mm²

[*] The minimum copper equivalent cross-sectional area is given by a copper bonding conductor of the tabulated cross-sectional area or a bonding conductor of another metal affording equivalent conductance.

Table 54.8 of BS 7671:2008 (2011)

3 What is the minimum csa required for a supplementary protective bonding conductor from Regulation 544.2.3, if not mechanically protected?

4 Show the equation used to obtain the minimum csa of a reduced size cpc, which does not comply with the tabulated values in Table 54.7 of BS 7671.

Isolation and isolation devices

Means of isolation should be provided as follows.

At the origin of the installation

A main-linked switch or circuit breaker should be provided as a means of isolation and of interrupting the supply on load. For single-phase household and similar supplies that may be operated by unskilled persons, a double-pole device must be used for both TT and TN systems.

For a three-phase supply to an installation forming part of a TT system, an isolator must interrupt the line and neutral conductors. In a TN-S or

TN-C-S system only the line conductors need be interrupted as the installation has a reliable connection to earth.

For every circuit

Other than at the origin of the installation, every circuit or group of circuits that may have to be isolated without interrupting the supply to other circuits should be provided with its own isolating device.

For every item of equipment

All fixed electrical equipment should be provided with a means of switching which can be used for the safety and maintenance of systems, from industrial production equipment to individual items, such as immersion heaters, hand dryers and showers.

For every motor

Every fixed electric motor should be provided with a readily accessible and easily operated device to switch off the motor and all associated equipment, including any automatic circuit breaker.

For every supply

All isolators must be labelled or identified, if it is not obvious which circuits they control.

Overcurrent protective devices

The designer of an installation will coordinate the loads, protective devices and sizing of cables. It is the inspector's job to verify that the correct protective devices have been installed as designed.

The following table explains the application of circuit breakers.

- The BS EN 60898 standard circuit breakers are Type B, C and D.
- The old versions of circuit breakers were the BS 3871 Type1, 2, 3 and 4. (These are now discontinued.)

Circuit-breaker type	Trip current (0.1 s to 5 s)	Application
1 B	2.7 to 4 I_n 3 to 5 I_n	Domestic and commercial installations having little or no switching surge
2 C 3	4 to 7 I_n 5 to 10 I_n 7 to 10 I_n	General use in commercial/industrial installations where the use of fluorescent lighting, small motors, etc., can produce switching surges that would operate a Type 1 or B circuit-breaker. Type C or 3 may be necessary in highly inductive circuits such as banks of fluorescent lighting
4 D	10 to 50 I_n 10 to 20 I_n	Not suitable for general use Suitable for transformers, X-ray machines, industrial welding equipment, etc., where high inrush currents may occur

NOTE: I_n is the nominal rating of the circuit-breaker.

Classification of circuit breakers (from the On-Site Guide, IET)

The cable current-carrying capacity must be checked. The cables connected to the protective devices, whether fuses or circuit breakers, must be checked for coordination.

Voltage drop calculation examples can be found on pages 110–113. Further coverage of coordination can be found on pages 64, 87, 96 and 129.

Understand the regulatory requirements and procedures for the safe testing and commissioning of electrical installations

APPROPRIATE TESTS AND SEQUENCE FOR ELECTRICAL INSTALLATIONS

The tests which need to be carried out during initial verification are listed below, in the correct sequence in accordance with BS 7671. Some of the tests will not be applicable to all electrical installations; however, the applicable tests must be carried out in order.

Regulation section 612 – prescribed tests

Dead tests

612.2.1   Continuity of protective conductors; including cpc, main and supplementary bonding conductors
612.2.2   Continuity of ring final circuit conductors
612.3   Insulation resistance
612.4.1   Protection by **SELV**
612.4.2   Protection by **PELV**
612.4.3   Protection by electrical separation
612.4.4   Functional extra-low voltage circuits
612.4.5   Basic protection by a barrier or enclosure provided during erection
612.5   Insulation resistance/impedance of floors and walls
612.6   Polarity
612.7   Earth electrode resistance (if applicable)

Live tests

612.7   Earth electrode resistance (if applicable)
612.8   Protection by automatic disconnection (verification in accordance with Chapter 41 of BS 7671)
612.9   Earth fault loop impedance
612.10   Additional protection
612.11   Prospective fault current
612.12   Check of phase sequence
612.13   Functional testing
612.14   Verification of volt drop

Assessment criteria

4.1 State the tests to be carried out on an electrical installation in accordance with the IET Wiring Regulations and IET Guidance Note 3

SELV

Separated extra-low voltage circuit.

PELV

Protective extra-low voltage circuit.

ACTIVITY

What would be the first four tests to be carried out on a standard domestic ring final circuit?

 SmartScreen Unit 307
Handout 17 and Worksheet 17

Regulation section 612 – prescribed tests for small industrial unit

If you consider the location plan of a small industrial unit, shown below and in Appendix 1 (page 227), you can see that some of the aforementioned tests will be necessary to prove that the electrical installation is safe to be put into use.

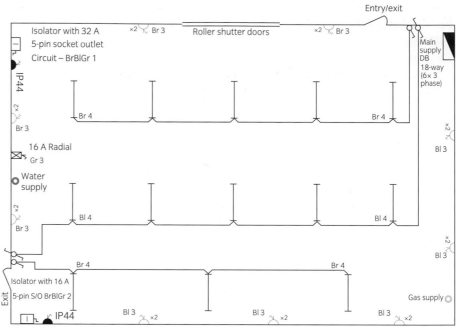

Small industrial unit (for larger version see Appendix 1)

shown below and in Appendix 1 (page 227)

(for larger version see Appendix 1)

The applicable tests for this small industrial unit are listed below.

Dead tests

612.2.1   Continuity of protective conductors; including main and supplementary equipotential bonding
612.2.2   Continuity of ring final circuit conductors
612.3   Insulation resistance
612.6   Polarity

Live tests

612.8   Protection by automatic disconnection (verification in accordance with Chapter 41 of BS 7671)
612.9   Earth fault loop impedance
612.10   Additional protection
612.11   Prospective fault current
612.12   Check of phase sequence
612.13   Functional testing
612.14   Verification of volt drop

As you progress through this unit:

- the tests given in the lists on page 44, relating to the location plan of the small industrial unit, will be calculated and test methods will be explained
- the tests not selected in the lists will also be explained with the aid of illustrations, test methods, calculations and table references.

Regulation section 612 – logging the test result data

The Generic Schedule of Test Results allows for the test results to be logged in a format that can be understood by all competent testers.

If you observe the progression through the Schedule of Test Results shown on page 46, all of the dead tests are followed by the live tests.

IDENTIFYING THE CORRECT INSTRUMENT FOR TESTS

Testing should be carried out in such a manner that no danger to person, livestock or property arises.

The multi-function test meter in the photograph shows that there are many different settings available for the various tests required by BS 7671. The meter may look complicated, but if you concentrate on one test at a time, you will understand how to use it. This section will simply outline the settings required for different tests and will give some sample test results.

Before you test

Before any tests are performed, you need to check the leads and probes to ensure that they are in good order. Methods of checking leads will be explained as you are shown each test in order.

The leads must be located correctly for testing. The individual meter settings for each test are briefly introduced on pages 47–48.

Form 4

GENERIC SCHEDULE OF TEST RESULTS

Form No: ..../4

DB reference no	....
Location	....
Z_s at DB (Ω)	....
I_{pf} at DB (kA)	....
Correct supply polarity confirmed	☐
Phase sequence confirmed (where appropriate)	☐

Details of circuits and/or installed equipment vulnerable to damage when testing ....

Details of test instruments used (state serial and/or asset numbers)
Continuity ....
Insulation resistance ....
Earth fault loop impedance ....
RCD ....
Earth electrode resistance ....

Tested by:
Name (Capitals) .... Date ....
Signature ....

Test results

Circuit details									Test results													
		Overcurrent device				Conductor details			Ring final circuit continuity (Ω)			Continuity (Ω) ($R_1 + R_2$) or R_2		Insulation Resistance (MΩ)		Polarity	Z_s (Ω)	RCD			Remarks (continue on a separate sheet if necessary)
Circuit number	Circuit Description	BS (EN)	type	rating (A)	breaking capacity (kA)	Reference Method	Live (mm²)	cpc (mm²)	r_1 (line)	r_n (neutral)	r_2 (cpc)	$(R_1 + R_2)$ *	R_2	Live-Live	Live-E			@ I$_{\Delta n}$ (ms)	@ 5I$_{\Delta n}$	Test button operation	
1	2	3	4	5	6	7	8	9	10	11	12	13	14	15	16	17	18	19	20	21	22

* Where there are no spurs connected to a ring final circuit this value is also the $(R_1 + R_2)$ of the circuit.

In this Schedule of Test Results, note that all of the dead tests are completed before the live tests (always check you are using the latest forms, as found on the IET website: http://electrical.theiet.org)

Setting up your test meter

Low-resistance ohmmeter

The tests carried out with this meter are:

- Continuity of protective conductors, including main and supplementary equipotential bonding.
- Continuity of ring final circuit conductors.
- Polarity.

The meter setting is in ohms (Ω), as shown in the photograph.

The test meter must be capable of supplying a no-load voltage of between 4 V and 24 V (d.c. or a.c.) and a short-circuit current of not less than 200 mA.

The measuring range should span 0.2 Ω to 2 Ω, with a resolution of at least 0.01 Ω for digital instruments.

Note that general purpose multi-meters are not capable of supplying these voltage and current parameters.

Insulation resistance tester

The tests carried out with this meter are:

- Insulation resistance testing.
- Separation of circuits, including
  - SELV or PELV
  - electrical separation.

The meter setting is in megohms (MΩ), as shown in the photograph.

The test meter must be capable of supplying an output test voltage of 250 V d.c., 500 V d.c. or 1000 V d.c., subject to the circuit to be tested.

The readings can range from 0.00 MΩ to to 2000 MΩ.

Earth fault loop impedance tester

The tests carried out with this meter are:

- Earth fault loop impedance.
- Prospective fault current.

The photograph shows the meter set to Loop – 20 Ω scale, which is usually used for TN systems and some TT systems.

An earth fault loop impedance tester with a resolution of 0.01 Ω should be adequate for circuits rated up to 50 A. Instruments conforming to BS EN 61557-3 will fulfil the above requirements.

Continuity testing – the left dial is set to the Ω scale and the right dial is set at 20 Ω

Insulation resistance testing – the left dial is set to 500 V and the right dial is set at 2000 MΩ

Earth fault loop impedance testing – this example shows the dial set at Loop – 20 Ω. Loop 200 Ω and 2000 Ω may need to be used when testing TT systems.

Prospective fault current testing – this example shows the dial set at PSC – 20 kA

RCD testing – this example shows the dial set at RCD – ×½ with the 30 mA setting displayed in the LCD

Earth fault loop impedance instruments may also offer additional facilities for deriving prospective fault current. The basic measuring principle is generally the same as for earth fault loop impedance testers.

The photograph shows the correct settings for deriving the prospective fault current.

Prospective fault current is measured in kA and can range from 0.3 kA to 16 kA.

The readings must be compared to the rated short-circuit capacity of the protective devices.

Residual current device (RCD) tester

The tests carried out with this meter are:

- Additional protection
- Residual current device operation.

Operation of RCDs must be checked to ensure that they are operating in accordance with the manufacturer's intended time limits. These results are given in milliseconds (ms). Methods and appropriate maximum trip times can be found in Section 11 of the IET On-Site Guide.

The photograph shows just one example of a typical setting for this test.

Note that earth electrode testing, polarity testing, phase rotation testing, functional testing and verification of voltage drop will be addressed as we progress through the test sequence.

ASSESSMENT GUIDANCE

All of these tests can be carried out by using a test meter as shown. However, when answering City & Guilds assessment questions, you will be expected to name the specific meter used in conjunction with the specific test, as shown in the activity on the right.

ACTIVITY

State the test instrument used and the test result unit of measurement expected for each of the tests listed below. For the example of RCD tests, the answer would be 'RCDs are tested with an RCD tester and the readings are in milliseconds (ms)'.

1 Continuity of protective conductors
'Continuity testing is carried out by using a …'

2 Insulation resistance testing
'Insulation resistance testing is …'

SAFE AND CORRECT USE OF INSTRUMENTS

When using test instruments, safety can be achieved by ensuring that the following points are followed.

Steps to follow when using instruments

1 Checking that the instruments being used conform to the appropriate British Standard safety specifications

- The basic instrument standard is BS EN 61557 'Electrical safety in low voltage distribution systems up to 1000 V a.c. and 1500 V d.c. Equipment for testing, measuring or monitoring of protective measures.'
- This standard includes performance requirements and requires compliance with BS EN 61010.

2 Instrument accuracy

- Instrument accuracy of 5% is usually adequate for testing.
- The expected accuracy for analogue meters is 2% of full-scale deflection.

3 Calibration

- Calibration must be carried out on each piece of test equipment in accordance with the manufacturer's recommendations.
- Regular checking may be carried out using known references.
- Usually, annual calibration suffices, unless the instrument is subject to excessive mechanical stresses.

4 Understanding the equipment

- You should understand the equipment to be used and its ranges.
- It is important, also, to understand the characteristics of the installation where the test instrument will be used.

5 Selecting and using the appropriate scales and settings

- It is essential that the correct scale and setting are selected for an instrument in a particular testing situation. As a vast range of instruments exists, always read the manufacturer's instructions before using an unfamiliar instrument. Further things to consider are:
  - When a meter displays a number 1 in the left of the display, it usually means that the instrument reading is over the range selected.
  - For continuity testing, the lowest ohms scale must be selected as values are usually low.

Assessment criteria

4.3 Specify the requirements for the safe and correct use of instruments to be used for testing and commissioning

ACTIVITY

What check should be carried out on a test instrument before and after use?

ASSESSMENT GUIDANCE

Unless the instrument is self-ranging, set it to the highest appropriate scale and then select progressively lower scales until the most accurate reading is achieved.

◻ Remember that the values shown while insulation resistance testing is taking place are in megohms (millions of ohms, MΩ).

◻ Many instruments feature a 'no-trip' function when testing earth fault loop impedance. This should only be used where an RCD is in the circuit. In all other cases the 'high' setting should be used as this is much more accurate.

◻ Manufacturers instructions must be followed when selecting the method of prospective fault current testing. These tests are explained on pages 100–102.

6 Checking test leads

■ Check that test leads, including any probes or clips, are in good order, are clean and have no cracked or broken insulation.

■ Where appropriate, the requirements of the Health and Safety Executive Guidance Note GS38 should be observed for test leads.

The photograph shows a test instrument with a range of test leads, probes and clips. These leads can be used for different tests.

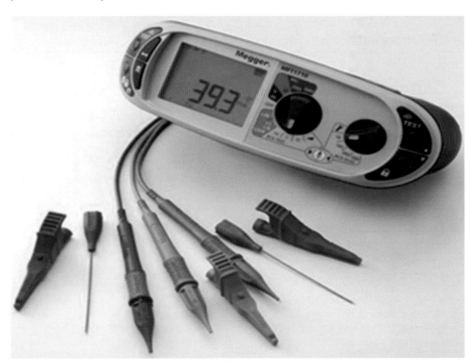

Multi-function test instrument showing a range of leads and probes (image courtesy of Megger)

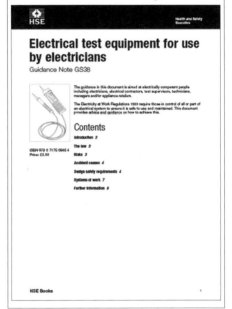

KEY POINT

Guidance Note GS38 gives guidelines on test equipment, leads and probes to be used when testing on or near live conductors. This can be freely viewed and downloaded from the Health and Safety Executive website: www.hse.gov.uk

WHY TEST RESULTS NEED TO COMPLY WITH STANDARDS

Test results must be recorded on the Schedule(s) of Test Results and compared with relevant criteria. For example, in order to verify disconnection times, the relevant criteria would be the design earth fault loop impedance values provided by the designer.

Each test is carried out for a reason as outlined in the table, with a very brief description of what could happen as a consequence of the conditions revealed by the test result.

Test	Risk(s) indicated by unsatisfactory result
Continuity of protective conductors, including main and supplementary bonding	Shock
Continuity of ring final circuit conductors	Overload, fire, shock
Insulation resistance	Fire, shock
Protection by SELV, PELV or by electrical separation	Shock
Protection by barriers or enclosures provided during erection	Shock
Insulation resistance of non-conducting floors and walls	Shock
Polarity	Shock, equipment damage
Earth electrode resistance	Shock
Protection by automatic disconnection of the supply	Shock
Earth fault loop impedance	Shock
Additional protection	Shock
Prospective fault current	Explosion, injury, fire damage
Phase sequence	Equipment damage, injury
Functional testing	Incorrect working of equipment, shock
Voltage drop	Equipment malfunction and damage

If any of the above tests prove to be unsatisfactory during an initial verification, the fault must be rectified before any more tests (in the proper order) are undertaken.

Assessment criteria

4.4 Explain why it is necessary for test results to comply with standard values and state the actions to take in the event of unsatisfactory results being obtained

ACTIVITY

Which BS 7671 appendix gives disconnection times for fuses and circuit breakers?

ACTIVITY

As you can see, shock is the greatest risk, but very few people are killed by electric shock each year. Can you think why this is?

TESTING IN THE EXACT ORDER SPECIFIED

The test sequence must be carried out in the exact order prescribed in Regulations 612.2 to 612.14 of BS 7671. Testing should stop if any circuit test result is not within the prescribed limits.

Some reasons for the specified order of testing are outlined below:

1 The test sequence has been developed to reduce the risk of electric shock during testing. All of the tests from Regulation 612.2 to 612.7 should be carried out before the installation is energised (Regulation 612.1).

2 If a test result within the prescribed test sequence fails to meet its minimum or maximum value, it may be dangerous or impractical to progress. For example:

- Impractical – if a continuity test for a lighting circuit records an open circuit, progressing to the insulation resistance test is impractical, because testing between conductors and earth is a worthless exercise.

- Dangerous – if a very low insulation resistance value was recorded during the testing of a circuit, this would indicate possible leakage currents. If this circuit was subsequently switched on for live testing, there could be fire or electric shock risk.

ACTIVITY

Except for short runs, it would be unusual for the earthing conductor to be visible throughout its length. What impact might this have on continuity testing?

CONTINUITY TESTING

Continuity testing of the earthing conductor

Note that secure isolation must be carried out before this test is done (secure isolation is covered on pages 4–6).

Reason for this test

This test must be carried out to ensure that the earthing conductor is not broken; it should connect the main earthing terminal to the means of earthing.

This test is not required if the earthing conductor can be *visually* inspected throughout its entire length. If at any point, the earthing conductor is concealed, this test is recommended.

Test meter

The test meter must be capable of supplying a no-load voltage between 4 V and 24 V d.c. or a.c. and a short-circuit current of not less than 200 mA.

The measuring range should span 0.01 Ω to 2 Ω, with a resolution of at least 0.01 Ω for digital instruments.

Note that general purpose multi-meters are not capable of supplying these voltage and current parameters.

Method

1   Select the correct meter – you will need a low-resistance ohmmeter.

2   Select the correct scale – the correct scale is ohms/continuity (see pages 49–50 for guidance on scales)

3   Check the leads for damage.

4   Insert the leads – choose the correct location on the meter (typically C and Ω, but this may differ for different instruments. Always consult the manufacturer's instructions.

5   Correctly connect the leads – the correct way to connect leads is explained in the diagram.

6   Zero or null the meter leads – the correct reading should be 0.00 Ω. If a particular test requires a longer lead or a link lead, this should be zeroed or nulled along with the meter leads.

You are now ready to carry out the continuity testing.

Carrying out the continuity testing of the earthing conductor

Look again at the location plan of the small industrial unit in Appendix 1 and the distribution and wiring information in Appendix 2. You need to identify the earthing conductor, the protective conductors and the circuits that require continuity testing.

The small industrial unit is fed from a TN-C-S supply. The main supply distribution board (DB) is situated 5 m away from the origin of the supply. The main three-phase (line conductors) and neutral supply tails have been calculated and designed to be 35 mm².

How to calculate the resistance

The cable cross-sectional area has been selected in accordance with Regulation 543.1.1 and Table 54.7 of BS 7671 (16 mm² earthing conductor).

The distance from the DB Main Earthing Terminal (MET) to the earthing conductor at the origin of the supply is 5 m.

Using the resistance value given in Appendix B (Table B1) of Guidance Note 3, a 16 mm² cable has a resistance of 1.15 mΩ/m.

Resistance of the cable is calculated as 5 m x 1.15 mΩ/m = 5.75 mΩ or 0.005 Ω.

Incorrect method of connecting leads

-ve   +ve

Current flow has to travel across the hinges

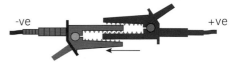

Correct method of connecting leads

-ve   +ve

Current flow is straight from clip to clip
Always connect leads correctly

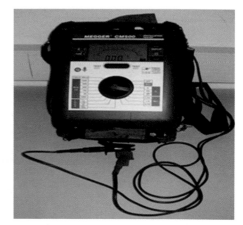

Multi-function test instrument with leads connected together (image courtesy of Megger)

Conductor csa mm²	Copper conductor resistance mΩ/m
4	4.61
10	1.83
16	1.15

Resistance values for cables (information extracted from Table B1 of Guidance Note 3)

How to test the resistance

For this test, the meter leads are not long enough.

- You must have a flexible earth extension reel – normally about 25 m length.
- Zero the test leads by connecting them to both ends of the extension reel.
- Connect one end of the extension lead to the earthing conductor, which needs to be disconnected from the MET (isolation of the supply to the building must be carried out before disconnecting the earthing conductor).
- Connect one crocodile clip to the earthing conductor at the origin of the supply and the other crocodile clip to the extension.

A 5 m length of 16 mm² should show on the meter (to two decimal places) as 0.01 Ω.

So, test result = 0.01 Ω for the earthing conductor.

The diagram below shows the earthing conductor and the main protective bonding conductor, with the earthing conductor 16 mm².

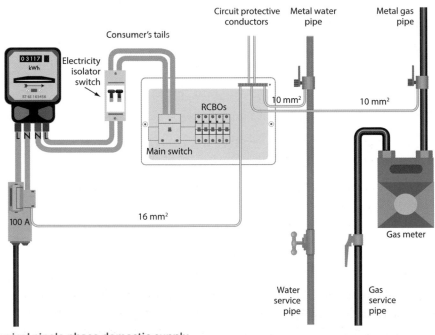

Typical single-phase domestic supply

Continuity testing of main and supplementary protective bonding conductors

Reason for this test

This test is carried out to ensure that all protective bonding conductors are continuous and have low resistance values. Note that secure isolation must be carried out before testing (secure isolation is covered on pages 4–6).

KEY POINT

The resistance of a cable is proportional to its length. If a cable is 10 m long and its resistance is 0.18 Ω, 50 m of the same cable will have a resistance of 0.90 Ω.

ACTIVITY

1 Measure the resistance of a partly used drum or a length of cable. Zero your meter and test the resistance.

2 Compare the measured resistance value to the estimated calculation value.

Test meter

The test meter must be capable of supplying a no-load voltage between 4 V and 24 V, d.c. or a.c., and a short-circuit current of not less than 200 mA.

The measuring range should span 0.2 Ω to 2 Ω, with a resolution of at least 0.01 Ω for digital instruments.

Note that general purpose multi-meters are not capable of supplying these voltage and current parameters.

Method

1 Select the correct meter – you will need a low-resistance ohmmeter.

2 Select the correct scale – the correct scale is ohms/continuity (see pages 49–50 for guidance on scales).

3 Check the leads for damage.

4 Insert leads – choose the correct location on the meter (typically C and Ω but this may differ between instruments). Always consult the manufacturer's instructions.

5 Correctly connect the leads – the correct way to connect leads is shown in the diagram.

6 Zero or null the meter leads – the correct reading should be 0.00 Ω. If a particular test requires a longer lead or a link lead, this should be zeroed or nulled along with the meter leads.

You are now ready to carry out the continuity testing.

Carrying out the continuity testing of the main protective bonding conductor

If you know the cross-sectional area (csa) of the earthing conductor, you can check the minimum csa of the main protective bonding conductors. The following information will allow you to assess the resistance of the main protective bonding conductors.

Look again at the location plan of the small industrial unit in Appendix 1. The small industrial unit is fed from a TN-C-S supply. The main supply DB is situated 5 m away from the origin of the supply. The main three-phase (line conductors) and neutral supply tails have been calculated and designed to be 35 mm². The earthing conductor is 16 mm².

How to calculate the resistance

You can calculate the value of resistance before carrying out continuity testing to give you an idea of the reading you should be obtaining.

From the location plan in Appendix 1, you can see that there are at least two main protective bonding conductors to be installed – the water utility supply and the gas utility supply.

Incorrect method of connecting leads

-ve   +ve

Current flow has to travel across the hinges

Correct method of connecting leads

-ve   +ve

Current flow is straight from clip to clip

Always connect leads correctly

Conductor csa mm²	Copper conductor resistance mΩ/m
4	4.61
10	1.83
16	1.15

Resistance values for cables (information extracted from Table B1 of Guidance Note 3)

The water supply main protective bonding conductor is 23 m long and selected in accordance with Regulation 544.1.1 and Table 54.8 of BS 7671 (10 mm²). The neutral of the supply is 35 mm².

In order to determine the resistance of the conductor at 20 °C, the following formula is used:

$$\frac{length \times m\Omega/m}{1000}$$

Therefore, the resistance would be:

$$\frac{23 \times 1.83}{1000} = 0.04 \ \Omega$$

How to test the resistance

To test the resistance:

- Zero the test leads by connecting them to both ends of the extension earth.

- Connect one end of the extension lead to the disconnected main protective bonding conductor for the water supply pipe at the MET (isolation of the supply to the building must be carried out before disconnecting the main protective bonding conductor from the MET).

- Connect one crocodile clip to the metal water supply connection and the other crocodile clip to the remaining end of the rolled out extension lead.

The illustration shows the earthing conductor (16 mm²) and the main protective bonding conductor (10 mm²).

ACTIVITY

How would you ensure good contact between the bonding conductor and water pipe?

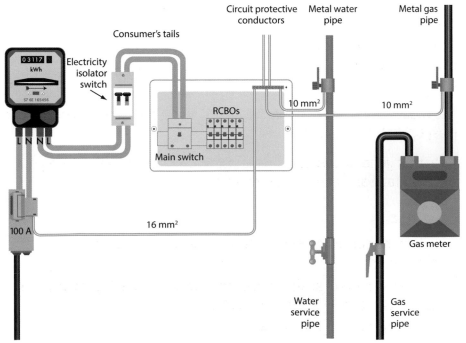

Typical single-phase domestic supply

A 23 m length of 10 mm^2 should show on the meter (to two decimal places) as 0.04 Ω, as shown in the example above.

So, test result for the water supply main protective bonding conductor is 0.04 Ω.

Carrying out the continuity testing of supplementary protective bonding conductors

Continuity testing of supplementary protective bonding conductors should be carried out to ensure that the bonding is continuous. It is suggested that the test results between conductive parts should be in the order of 0.05 Ω or less. Further details on specific results is given in BS 7671.

In a room containing a bath or shower, tests can be carried out with a low-resistance ohmmeter to prove bonding between, for example, hot and cold water, central heating, heated towel rail, lighting circuit and shower.

ACTIVITY

The main protective bonding conductor to the gas utility supply is 10 mm^2 csa and 13 m in length. Calculate the resistance of the cable, showing the expected meter reading to two decimal places.

Continuity of cpc for lighting circuits and radial circuits

Note that secure isolation must be carried out before this test is done (secure isolation is covered on pages 4–6).

Reason for this test

This test is carried out to ensure that a cpc is present and effectively connected to *all* points in the circuit, including all switches, sockets, luminaires and outlets.

ASSESSMENT GUIDANCE

Remember, the $R_1 + R_2$ test for continuity will also cover polarity testing. Don't waste time by doing the test again.

Test meter

The test meter must be capable of supplying a no-load voltage between 4 V and 24 V d.c. or a.c. and a short-circuit current of not less than 200 mA.

The measuring range should span 0.01 Ω to 2 Ω, with a resolution of at least 0.01 Ω for digital instruments.

Note that general purpose multi-meters are not capable of supplying these voltage and current parameters.

Method

1　Select the correct meter – you will need a low-resistance ohmmeter.

2　Select the correct scale – the correct scale is ohms/continuity (see pages 49–50 for guidance on scales).

3　Check the leads for damage.

Incorrect method of connecting leads

Current flow has to travel across the hinges

Correct method of connecting leads

Current flow is straight from clip to clip

Always connect leads correctly

4 Insert leads – choose the correct location on the meter (typically C and Ω but this may differ between instruments). Always consult the manufacturer's instructions.

5 Correctly connect the leads – the correct way to connect leads is shown in the diagram.

6 Zero or null the meter leads – the correct reading should be 0.00 Ω. This test may require a link lead; this should be zeroed or nulled along with the meter leads.

You are now ready to carry out the continuity testing; in this example, for a lighting circuit and a radial circuit.

Carrying out the continuity testing for a lighting circuit

There are two methods of testing and recording results for the continuity of a lighting circuit. Test methods 1 and 2 are illustrated below and opposite.

Test method 1 – $R_1 + R_2$

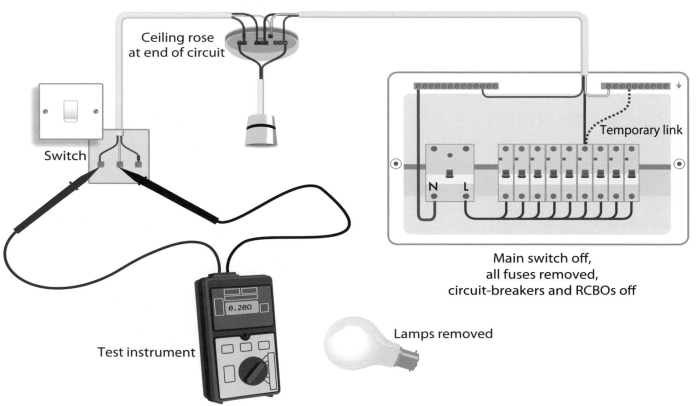

Continuity testing at a light switch using test method 1

Test method 1 involves temporarily linking between the line and cpc, and testing at *all* relevant points of the circuit; that is at switches and luminaires. The highest $R_1 + R_2$ value is recorded – usually at the furthest point away from the DB.

Look again at the location plan and the circuit information relating to the small industrial unit in Appendix 1.

If you use the data for the lighting circuit fed from 'Br-4', you have 37 m of 1.5 mm² line and 1.5 mm² cpc. Using the table below, the expected $R_1 + R_2$ value will be:

$$\frac{37 \times 24.2}{1000} = 0.89 \ \Omega \ \text{(at 20 °C)}$$

Line conductor csa mm² (R_1)	Circuit protective conductor csa mm² (R_2)	Copper conductor resistance mΩ/m at 20 °C
1.5	–	12.10
1.5	1.5	24.20
2.5	–	7.41
2.5	2.5	14.82
4.0	1.5	16.71

Resistance values for cables (information extracted from Table B1 of Guidance Note 3)

Test method 2 (wandering lead test for R_2 resistance)

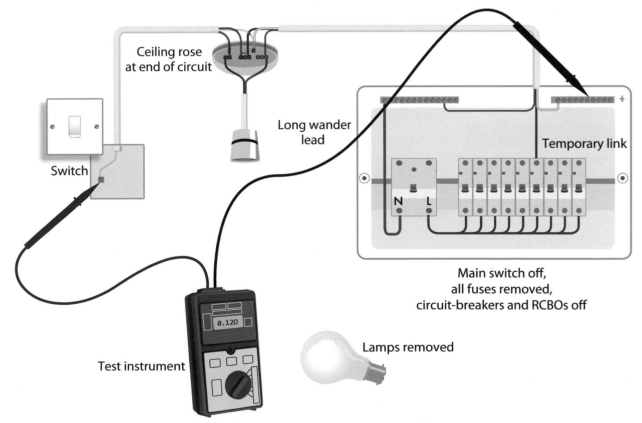

Continuity testing at a light switch using test method 2

Test method 2 involves the tester applying the long lead and zeroing the meter, as shown when testing the main protective bonding conductor, on page 55.

This method is rarely used to verify a radial circuit as obtaining a value of $R_1 + R_2$ is a minimum for initial verification, whereas a value of R_2 only proves the presence of an earth or cpc. A value of $R_1 + R_2$ will be used later when obtaining values of earth fault loop impedance.

Look again at the location plan and the circuit information relating to the small industrial unit in Appendix 1.

If you assume that the cpc is 37 m in length, the R_2 value will be 0.45 Ω.

Carrying out the continuity testing for a radial circuit (eg instantaneous water heater)

Using the location plan and circuit information relating to the small industrial unit in Appendix 1, you can see that the 3 kW instantaneous water heater is fed from circuit 'Gr-3' which is protected by a B-type 16 A circuit breaker.

All conductors have a csa of 2.5 mm^2 and the length of run 22 m.

Before testing, you should have some idea of the expected test results. From the information given, you can calculate the approximate resistance of the line conductor (R_1) and the cpc (R_2) at an ambient temperature of 20 °C.

Calculation of the resistance

$$\frac{22 \times 14.82}{1000} = 0.33 \, \Omega$$

Testing the resistance

To test the resistance:

- Zero the test leads, including any temporary link lead using the ohms scale (continuity).

- Using test method 1 shown for the lighting circuit above, temporarily link between the line and cpc at the DB using a temporary link lead.

- With the instantaneous water heater cpc disconnected (to avoid parallel earth paths), test between the line conductor and the cpc.

Expected test results for $R_1 + R_2$ of the instantaneous water heater = 0.33 Ω.

Continuity testing for three-phase socket outlets and fixed equipment

1 Using test method 1 shown for the lighting circuit above, temporarily link between:

   a) line 1 and cpc at the DB

   b) line 2 and cpc at the DB

   c) line 3 and cpc at the DB.

2 At the socket outlet, test between line and cpc for each line. Record all three test results (they should all be similar values). Some parts of the industry suggest that only the highest reading should be recorded as the $R_1 + R_2$ for a three-phase circuit, but it is generally best to record all three results so patterns of failure or deterioration can be detected during periodic inspections.

Continuity of ring final circuit conductors

Reason for this test

This test is carried out to ensure that the ring final circuit actually forms a ring and that any spurs off the ring are adequately protected. This test process will also confirm polarity of the ring final circuit, as well as obtaining the $R_1 + R_2$ for the ring – confirming continuity of cpc.

Test meter

Note that secure isolation must be carried out before this test is done (secure isolation is covered on pages 4–6).

The test meter must be capable of supplying a no-load voltage between 4 V and 24 V d.c. or a.c. and a short-circuit current of not less than 200 mA.

The measuring range should span 0.01 Ω to 2 Ω, with a resolution of at least 0.01 Ω for digital instruments.

Note that general purpose multi-meters are not capable of supplying these voltage and current parameters.

Method

1 Select the correct meter – you will need a low-resistance ohmmeter.

2 Select the correct scale – the correct scale is ohms/continuity (see pages 49–50 for guidance on scales).

3 Check the leads for damage.

4 Insert leads – choose the correct location on the meter (typically C and Ω but this may differ between instruments). Always consult the manufacturer's instructions.

KEY POINT

Always do a quick calculation before testing. This will allow you to estimate the correct meter reading.

ACTIVITY

Look again at the location plan and the circuit information relating to the small industrial unit in Appendix 1.

1 Calculate the $R_1 + R_2$ of the 'Bl-3' radial circuit for sockets outlets.

2 Calculate the $R_1 + R_2$ values for the 16 A three-phase socket outlet fed from circuit 'Br/Bl/Gr-2'.

ASSESSMENT GUIDANCE

Exam questions often ask learners to state the first three parts of the ring final circuit test. Learners often mistakenly name the resistance measurement of line, neutral and cpc conductors. This is incorrect. These comprise just the first stage.

Incorrect method of connecting leads

Current flow has to travel across the hinges

Correct method of connecting leads

Current flow is straight from clip to clip

Always connect leads correctly

5 Correctly connect the leads – the correct way to connect leads is shown in the diagram.

6 Zero or null the meter leads – the correct reading should be 0.00 Ω. If a particular test requires a longer lead or a link lead, this should be zeroed or nulled along with the meter leads.

You are now ready to carry out the continuity testing.

Carrying out continuity testing of the ring final circuit

There are three steps that you need to take to ensure that the ring final circuit has been installed correctly. These are outlined below and in the following pages.

Look again at the location plan and the circuit information relating to the small industrial unit in Appendix 1.

If you use the data for the ring final circuit, supplied from 'Br-3', you have a total ring distance of 48 m. There are no spurs off the ring final circuit.

The line and neutral have a csa of $2.5\,mm^2$ and the cpc has a csa of $1.5\,mm^2$.

<table>
<tr><th>Line conductor csa mm² (R_1)</th><th>Circuit protective conductor csa mm² (R_2)</th><th>Copper conductor resistance mΩ/m at 20 °C</th></tr>
<tr><td>2.5</td><td>–</td><td>7.41</td></tr>
<tr><td></td><td>1.5</td><td>12.10</td></tr>
<tr><td>2.5</td><td>1.5</td><td>19.51</td></tr>
<tr><td>2.5</td><td>2.5</td><td>14.82</td></tr>
</table>

Resistance values for cables (information extracted from Table B1 of Guidance Note 3)

ACTIVITY

What effect would there be on the readings for the ring final circuit if a $2.5\,mm^2$ cpc were used instead of $1.5\,mm^2$ cpc?

Step 1
Calculate the 'end-to-end' resistance of the line, neutral and cpc.

$$\text{Line } (r_1) = \frac{48 \times 7.41}{1000} = 0.35\ \Omega$$

$$\text{Neutral } (r_n) = \frac{48 \times 7.41}{1000} = 0.35\ \Omega$$

$$\text{cpc } (r_2) = \frac{48 \times 12.10}{1000} = 0.58\ \Omega$$

Note the test results. You will need the test results when calculating Steps 2 and 3.

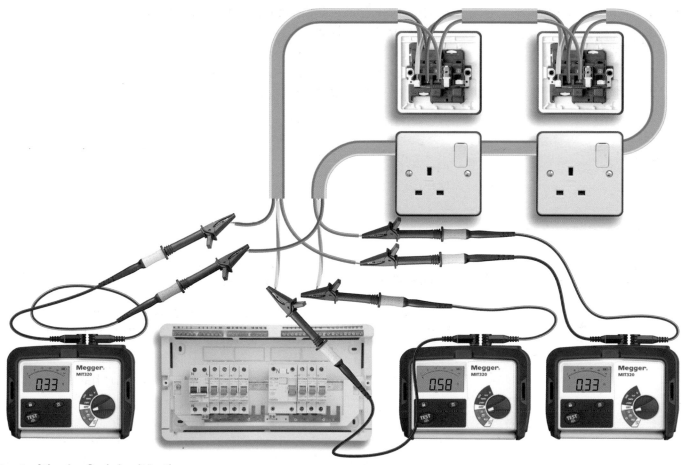

Step 1 of the ring final circuit testing

As shown in the diagram, measure the end-to-end resistance values for the line, neutral and cpc loops. These values are recorded in the Schedule of Test Results as r_1 (line loop), r_n (neutral loop) and r_2 (cpc loop). These test results may vary slightly from the calculated values due to **contact resistance** or cable length estimation.

Step 2

At the DB, cross connect the line and neutral. It is important to ensure that the line and neutral are cross connected as shown in the diagram. Incorrect cross connection will result in false readings. If readings taken at the socket outlets between line and neutral increase greatly towards the centre of the ring and then decrease again, the cross connections are incorrect.

Contact resistance

Sometimes, a meter may give a reading in which the value does not stabilise or which is quite different from the calculated value. In this situation, reverse the test meter lead connections and re-test. This may give more accurate values.

KEY POINT

If the DB is crowded, this part of the test may be carried out at a convenient socket outlet. If this is done, care must be taken to ensure that the selected socket outlet is isolated.

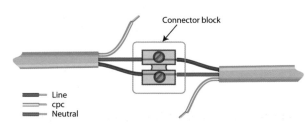

Step 2 of the ring final circuit testing

Calculation: the easiest way to remember the calculation is to add together the resistance of each complete ring from Step 1 above and divide by 4.

$$\frac{r_1 + r_n}{4} \text{ or } \frac{0.33\,\Omega + 0.33\,\Omega}{4} = 0.16\,\Omega$$

So, the reading at each socket outlet should be approximately 0.16 Ω.

Method of testing: continuity of line and neutral connection at each of the socket outlets on the ring final circuit should be tested. All results should be the same, however slight deviations do occur due to contact resistance.

The use of special adaptors is recommended when testing at the socket outlets as this allows readings to be obtained from the front plate of the socket outlet, instead of accessing the connections behind.

Typical test results
Socket outlet 1 = 0.17 Ω
Socket outlet 2 = 0.16 Ω
Socket outlet 3 = 0.18 Ω
Socket outlet 4 = 0.17 Ω

Note that these tests and results are important to ensure that there are no interconnections or extended spurs off spurs on this ring final circuit.

If a reading is substantially higher than expected, the circuit must be investigated further to ensure coordination between the current-carrying capacity of the cables and the protective devices. Before any investigation is carried out for inconsistent readings, obtain values as outlined in Step 3 – these values can give a good indication of what is causing the problem.

Step 3
At the DB, cross connect the line and cpc. It is important to ensure that the line and cpc are cross connected as shown below.

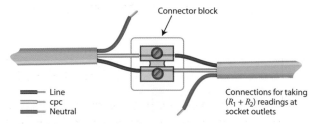

Step 3 of the ring final circuit testing

ACTIVITY

Why is it not necessary to test the cross connected N–E loop in the ring final circuit test after the L–N and L–E test?

Calculation: again, the easiest way to remember the calculation is to add together the resistance of each complete ring from Step 1 above and divide by 4.

$$\frac{r_1 + r_2}{4} \text{ or } \frac{0.33\,\Omega + 0.58\,\Omega}{4} = 0.23\,\Omega$$

So, the reading at each socket outlet should be approximately 0.23 Ω.

Method of testing: continuity of line and cpc connection at each of the socket outlets on the ring should be tested. All results should be the same, however slight deviations can occur due to contact resistance.

Typical test results
Socket outlet 1 = 0.24 Ω
Socket outlet 2 = 0.23 Ω
Socket outlet 3 = 0.25 Ω
Socket outlet 4 = 0.23 Ω

Note that:

■ The highest of these values will be inserted onto the Generic Schedule of Test Results as the $R_1 + R_2$ value and further evaluation will be required when assessing the earth fault loop impedance of the circuit.

$$R_1 + R_2 = 0.25\,\Omega$$

■ If there is a spur off the ring final circuit, the reading at that spur could be higher than those at socket outlets on the ring. The highest reading obtained will be recorded as the $R_1 + R_2$ value.

THEORY RELATED TO INSULATION RESISTANCE TESTING

The minimum acceptable values of insulation resistance of a circuit are shown in the table on page 69, taken from Table 61 of BS 7671. The purpose of insulation around a conductor is to contain the current in the conductor; the lower the insulation resistance, the greater the risk of current leaking between conductors or between conductors and earth.

Why you need to test for insulation resistance

The purpose of this test is to verify that the insulation of conductors provides adequate insulation, is not damaged and that live conductors or protective conductors are not short-circuited or leaking overcurrent.

KEY POINT

Always do a quick calculation before testing. This will allow you to estimate the correct meter readings for each step. Also, keep checking the instrument is zeroed or nulled, especially if special adaptors are used which may affect results.

ACTIVITY

Practically test a ring final circuit by using the methods explained here.

Assessment criteria

4.8 State the effects that cables connected in parallel and variations in cable length can have on insulation resistance values

4.9 Interpret and apply the procedures for completing insulation resistance testing

SmartScreen Unit 307
Handout 24

What is insulation resistance?

Insulation resistance is the resistance of the insulation between conductors. The higher the resistance, the better the current is contained in the core of the cable. If insulation resistance is low, current will leak between conductors and to earth, giving a risk of electric shock or fire.

Many years ago, rubber-insulated cables were installed in buildings. Unfortunately, as some of these cables deteriorated with age, they started to crack and insulation failure occurred. Long cable runs gave low values of insulation resistance, and multiple circuits from a common source, such as a distribution board, were used.

For example, if an insulation resistance test is carried out on a 50 m length of old rubber cable (between line and neutral) and the reading is 20 MΩ, a 100 m length of the same cable would give a reading of 10 MΩ. Why? Because you doubled the length; theoretically, you have put two 50 m lengths of cable in parallel.

Two resistances in parallel:

$$\frac{1}{R_1} + \frac{1}{R_2} = \frac{1}{R_t}$$

therefore,

$$\frac{1}{20} + \frac{1}{20} = \frac{1}{R_t}$$

so,

$$\frac{1}{20} + \frac{1}{20} = 0.05 + 0.05 = 0.1$$

and, finally

$$\frac{1}{0.1} = 10 \, (MΩ)$$

These calculations can be performed on a calculator using the x^{-1} feature.

Try this: [20] $[x^{-1}]$ [+] [20] $[x^{-1}]$ [=] $[x^{-1}]$ [=] and the answer displayed should be 10.

A calculation to illustrate circuit readings accumulated in parallel

Here are five individual final circuit test results, measured between line and cpc for each circuit:

- Circuit 1 = 25 MΩ (R_1)
- Circuit 2 = 10 MΩ (R_2)
- Circuit 3 = 16 MΩ (R_3)

ACTIVITY

This calculation can also be done as product over sum. What would be the total insulation resistance of two cables in parallel having individual insulation test values of 16 MΩ and 50 MΩ?

KEY POINT

The insulation resistance value decreases as the conductor length is increased.

ASSESSMENT GUIDANCE

You could be given the insulation resistance values of three, four or even five circuits which have been individually tested from the same distribution board (DB). You may be asked to calculate the cumulative value if all of the circuits are tested together.

- Circuit 4 = 6 MΩ (R_4)
- Circuit 5 = 40 MΩ (R_5)

Example questions

1   Determine the total value of these results in parallel.

2   Is the calculated value acceptable in accordance with the table taken from Table 61 of BS 7671?

Answer to example question 1

$$\frac{1}{R_t} = \frac{1}{R_1} + \frac{1}{R_2} + \frac{1}{R_3} + \frac{1}{R_4} + \frac{1}{R_5} = \frac{1}{25} + \frac{1}{10} + \frac{1}{16} + \frac{1}{6} + \frac{1}{40}$$

Now, with a calculator, change all the fractions to decimals (to two decimal points), ie: $1 \div 25 = 0.04$

$0.04 + 0.10 + 0.06 + 0.17 + 0.03 = 0.40$ total

To find the total value of these resistances in parallel, you must finally divide the total into 1.

Therefore, $\dfrac{1}{0.40} = 2.5$ MΩ

Answer to example question 2

So, is 2.5 MΩ an acceptable value by measurement and calculation? Yes it is – the minimum value given in Table 61 of BS 7671 = 1.0 MΩ.

> **KEY POINT**
>
> The calculated value for the total resistance will always be less than the lowest resistance in the set (in this case 6 MΩ).

Expected insulation resistance readings

The insulation resistance readings expected for a new installation are very high, for example greater than 200 MΩ for each circuit. For an older circuit that may have damage, dust or dampness in the system, readings can be considerably less than 200 MΩ.

> **ASSESSMENT GUIDANCE**
>
> There is a good reason for doing the continuity of protective conductor test before the insulation resistance test. If there is no cpc continuity, then the insulation test is worthless.

Whatever the readings, they must not be less than the minimum values stated Table 61. If the insulation resistance value obtained for a circuit was less than 1.0 MΩ, there could be excessive leakage of current occurring within the wiring system.

If you get a reading of 0.25 MΩ at 500 V between live conductors and earth, the resultant current flowing to earth in this circuit must be:

$$\frac{500\text{ V}}{250\,000\,\Omega} = 0.002\text{ A (or 2 mA)}$$

This leakage current could be sufficient to cause electric shock, high touch voltages or combustion, due to tracking across between conductors.

SmartScreen Unit 307
Worksheet 25

Practical guidance to insulation resistance testing

Note that secure isolation must be carried out before this test is done (secure isolation is covered on pages 4–6).

Reason for this test

This test is carried out to ensure that the insulation between conductors has a high value of resistance.

Test meter

The test meter must be capable of supplying an output test voltage of 250 V d.c., 500 V d.c. or 1000 V d.c., subject to the circuit to be tested.

See Table 61, from BS 7671, on page 69.

Method

1   Select the correct meter – you will need an insulation resistance tester. As the voltages used by the meter are typically 250 V, 500 V or 1000 V, the requirements of Guidance Note GS38 apply.

2   Select the correct scale – the correct scale is megohms (MΩ).

3   Check the leads for damage – clips and probes can be used when carrying out these tests.

4   Insert leads – choose the correct location on the meter (typically C and Ω, but this may differ between instruments). Always consult the manufacturer's instructions.

5   Choose the voltage – the voltage must be appropriate for the circuit under test. See Table 61 from BS 7671, on page 69.

6   Check the meter and leads.

- open circuit reading should be >200 MΩ
- connected together reading should be 0.00 MΩ

Note that the open circuit reading may differ from that shown here and is dependent on the manufactured parameters for the test meter used. Examples range from >20 MΩ to >1000 MΩ.

Why test insulation resistance?

The purpose of the insulation test is to verify that the insulation of conductors provides adequate insulation, is not damaged and that live conductors or protective conductors are not short-circuited or leaking overcurrent that could give rise to fire risk or electric shock.

The table below shows the minimum reading for an installation or a single circuit.

TABLE 61 – Minimum values of insulation resistance

Circuit nominal voltage (V)	Test voltage d.c. (V)	Minimum insulation resistance (MΩ)
SELV and PELV	250	0.5
Up to and including 500 V with the exception of the above systems	500	1.0
Above 500 V	1000	1.0

Minimum values of insulation resistance (from BS 7671:2008 (2011))

You are now ready to prepare and carry out the insulation resistance testing, using the methods outlined below and in the following pages.

Before testing

1 Ensure that these items are disconnected to reduce risk of damage:

- any electronic equipment such as dimmer switches, controllers and timers

- surge protective devices (SPDs)

- residual current devices (RCDs) and residual current operated circuit breakers with overcurrent protection (RCBOs)

2 Carry out these actions to enable full testing of the wiring system and to reduce the possibility of incorrect readings:

- remove all lamps

- connect all protective conductors

- disconnect any current-using equipment

- check all fuses are in place, and circuit breakers and switches closed

- all two-way and intermediate switching is operated during testing, so that the strappers are all tested.

Note that, if current-using equipment cannot be disconnected or lamps removed, switchgear supplying these items should be in the open position. Where an initial verification is undertaken, insulation resistance testing may be carried out during and on completion of the work. This will ensure all parts of a circuit are subjected to insulation resistance tests.

ACTIVITY

It is necessary to disconnect a dimmer before carrying out an insulation resistance test. What further action is required before the test is carried out?

ASSESSMENT GUIDANCE

Many people think that neon indicator lights will be damaged by an IR test. This is incorrect, they just glow. This will give a false insulation resistance test reading between line and neutral.

Single-phase testing methods

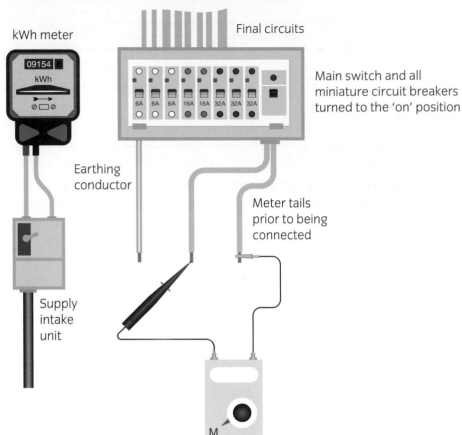

Line to neutral test being carried out

Note that all earthing and bonding *must* be connected for this test.

When testing the full installation or individual circuits, three tests are used:

- Test 1 – line to neutral (live to live)
- Test 2 – line to earth (live to earth)*
- Test 3 – neutral to earth (live to earth)*.

*The lowest value of the two tests is recorded.

An alternative test is line and neutral (together) to earth. This is used for circuits or equipment that are vulnerable to the test voltage, such as capacitors in fluorescent luminaires. All earths must be connected to the main earthing terminal (MET) for this test.

Continuity	Insulation resistance			
$(R_1 + R_2)^* \, \Omega$	R_2^*	ring	live/live MΩ	live/earth MΩ
			> 200	> 200

Extract from the Generic Schedule of Test Results showing the typical insulation test result of a new circuit

*The lowest value of the two tests.

If the readings for the tests of each circuit are greater than 200 MΩ, that information needs to be recorded on the Generic Schedule of Test Results. Remember that most meters will only read values up to 200 MΩ. If the meter indicates an over-range value, this is recorded as >200 MΩ.

Three-phase, neutral and earth (six-test sequence)

The six-test sequence is outlined below and on the next page. Note that Tests 5 and 6 can be carried out together if connected as shown in the diagram.

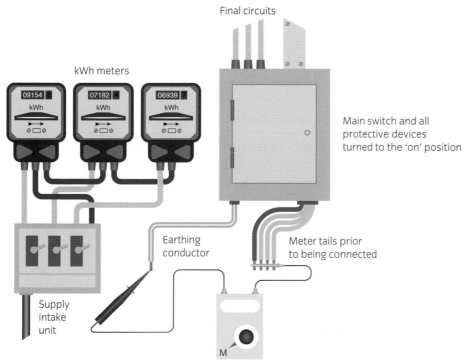

Tests 5 and 6 being carried out together – L1, L2, L3 + neutral (connected together) to earth

Note: all earthing and bonding *must* be connected for this test.

KEY POINT

Always ensure that precautions are taken to avoid damage to equipment and false readings.

ACTIVITY

Practise insulation resistance testing on single- and three-phase test boards at your college or training centre. Various types of boards and switchgear should be made available. Remember that these tests must only be carried out under supervision. Ensure that the test boards are fully isolated and secured before testing commences.

SELV

Separated extra-low voltage circuit.

PELV

Protective extra-low voltage circuit.

Test 1	L1 to L2	The lowest value of these tests is recorded as 'between live conductors'
Test 2	L1 to L3	
Test 3	L2 to L3	
Test 4	L1 + L2 + L3 (connected together) to neutral	
Test 5	L1 + L2 + L3 (connected together) to earth	The lowest value of these tests is recorded as 'between live conductors and earth'
Test 6	neutral to earth	

Three-phase and earth (four-test sequence)

Tests 1, 2, 3 and 5 will apply if no neutral is present, as in the case of an armoured cable supplying a three-phase motor.

Additional insulation resistance tests

Protection by SELV, PELV and electrical separation

In certain situations within an installation, electric shock protection under fault conditions is provided by **SELV** and **PELV** or electrical separation instead of automatic disconnection of supply (ADS). Where this exists, a test for simple separation is required between the low voltage circuit supplying the separated circuit and the outgoing circuit, as well as the outgoing circuit and earth.

This is done to ensure that there is no electrical connection between the separated circuit and the rest of the installation. It is done using an insulation resistance tester. If there was an electrical connection, the protective measure would be lost.

The test is carried out with the transformer that provides separation disconnected and the test instrument set to 500 V d.c., regardless of the circuit voltage. The instrument is connected between the supply circuit live conductors and the separated circuit live conductors. The result obtained must be greater than 1 MΩ.

A further test is then carried out between the live conductors of the separated circuit and earth. This again is carried out at 500 V d.c. and the result must be greater than 1 MΩ.

THE IMPORTANCE OF VERIFYING POLARITY

Note that verification of polarity must be carried out on all circuits.

If polarity is not correctly determined there may be a risk of electric shock during maintenance procedures. For example, if you isolate or switch the neutral of a circuit via a single-pole circuit breaker or switch, it would appear that the circuit is dead. This would not be the case, as live conductors and connections would be present at fixed equipment, sockets and switches – this would be very dangerous.

There are three recognised methods of evaluation. All three methods have their advantages and possible dangers, if they are not carried out correctly.

The methods are:

- polarity by visual inspection
- polarity by continuity testing
- live testing for polarity.

Polarity by visual inspection

By using your knowledge and sight, correct termination of cables relating to core colours can be established. It is essential that polarity is checked visually during the process of installation, especially in cases where checking by testing is impractical.

You can inspect:

- line conductors, which must be connected to circuit breakers, fuses, single pole switches and single-pole controls
- as highlighted in both BS 7671 and Guidance Note 3, that the centre contact of all screw-type lampholders is connected to the line conductor (except for E14 and E27 lampholders which have been manufactured to BS EN 60238; these lampholders have been manufactured to reduce the risk of shock whilst inserting lamps)
- correct connection of fittings and accessories
- all fixed equipment, such as supply tails, lights, socket outlets, heaters, motors and boilers.

Assessment criteria

4.10 Explain why it is necessary to verify polarity

4.11 Interpret and apply the procedures for testing to identify correct polarity

ASSESSMENT GUIDANCE

Remember that different makes of socket outlets may have their L and N terminals in different positions. Do not assume in your testing that all are the same.

ACTIVITY

Polarity tests are not required at a bayonet cap lampholder. Where would the nearest test point be?

Polarity by continuity testing

You will need to use a low-resistance ohmmeter for this test.

When you continuity test radial and ring final circuits, part of the process is to test and visually inspect the polarity of fixed equipment and socket outlets.

The diagram reinforces the polarity of the switch as being supplied by the line conductor, due to the line and cpc being linked in the consumer unit.

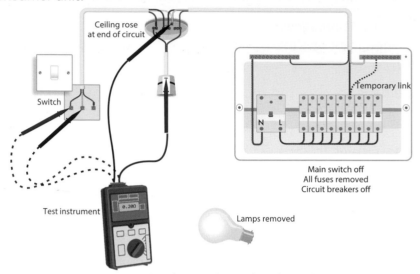

Main switch off
All fuses removed
Circuit breakers off

Lamps removed

Note: the test may be carried out either at lighting points or switches

Polarity test on lighting circuit

Note that when testing is complete, the temporary link must be removed.

All radial circuits that are correctly tested for continuity using the $R_1 + R_2$ method, are normally visually inspected, tested and verified for polarity at the same time. All socket outlets on ring final circuits will have been tested for polarity during the three steps of testing. Three-phase circuits are tested for continuity and, therefore, polarity is also tested.

For further information and safety procedures regarding continuity testing, refer to radial and ring final circuit testing on pages 52–65.

Live testing for polarity

You will need to use voltage indicators and earth fault loop impedance test meters. All test meters used for testing 'live circuits' must comply with Guidance Note GS38.

You can test these features for polarity when the system is energised:

- supply tails and earthing conductor
- distribution boards and consumer units
- socket outlets.

An ES E27 lampholder test adaptor can be used during continuity testing for polarity

ACTIVITY

State, for **each** of the following stages, **three** items of equipment which can be checked for polarity.

1 inspecting an installation
2 continuity testing
3 live testing

METHODS OF DETERMINING EARTH FAULT LOOP PATHS

There are three types of system earthing arrangements to consider:

- **TN-S**
- **TN-C-S**
- **TT**.

There are three methods of determining the external earth fault loop impedance (Z_e).

1 Measurement: by testing at the origin of the installation – this is the most common and accurate method used.

2 Enquiry: such as by contacting the electricity distributor, who will usually quote the maximum possible Z_e value in accordance with the Electricity, Safety, Quality and Continuity Regulations 2002; you will require permission to use this system.

3 Calculation: this is the most complex method and can only be used if the designer knows all of the relevant impedances and resistances of the system to be tested.

TN-S system and earth fault loop impedance

This section concentrates on the **TN-S** system. TN-S systems may be single-phase or three-phase.

Three-phase supply for a TN-S system

The three-phase supply for a TN-S installation is shown in the diagram. The electricity distributor will supply the consumer with an earth, which is separated throughout the system, but connected to the neutral at centre point of the supply transformer.

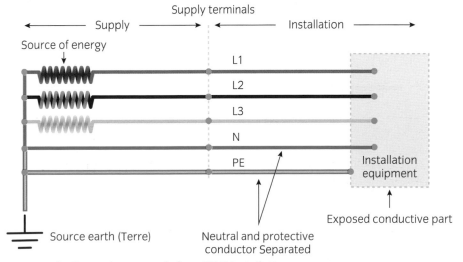

Diagram of a three-phase supply for a TN-S installation

Single phase TN-S system

The diagram shows a practical arrangement for a single-phase TN-S system. It is typical of a domestic (or similar) installation.

In both single- and three-phase supplies, a TN-S system will normally use the sheath of the supplier's cable as a means of earthing the installation. From this sheath, the earthing conductor will connect to the installation's main earthing terminal (MET).

An enquiry to an electricity distributor regarding the characteristics of a domestic supply will result in the following information being provided:

- external earth fault loop impedance (Z_e) = 0.8 Ω, the maximum for separate earth supplies in TN-S systems
- the Z_e value in practice will normally be considerably less than 0.8 Ω.

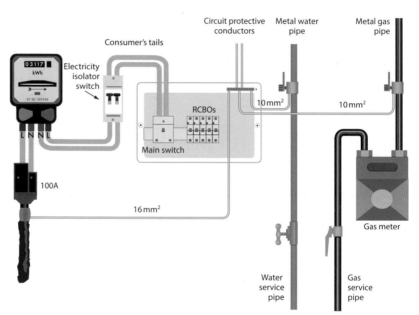

Practical arrangement for a single-phase TN-S system

The external earth fault loop impedance (Z_e) in ohms is a result of the impedance of the supplier's line conductor, the transformer and the separate earth added together.

KEY POINT

Three methods of determining the external earth fault loop impedance are by measurement, enquiry or calculation.

ACTIVITY

Redraw the simple TN-S system diagram and illustrate how you would attach the probes or leads of a two-lead test meter to the supply to test the earth fault loop impedance (Z_e) at the origin. Place as much information on the diagram as possible and show the maximum expected reading on the meter.

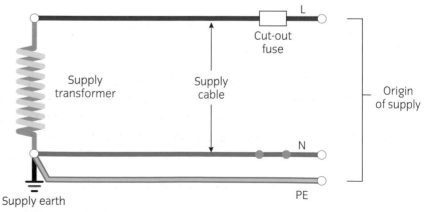

Simple representation for a TN-S system

TN-C-S system and earth fault loop impedance

This section concentrates on the **TN-C-S** system (**PME**).

There are three methods of determining the earth fault loop impedance (Z_e).

1 Measurement: by testing at the origin of the installation – this is the most common and accurate method used.

2 Enquiry: such as by contacting the electricity distributor, who will usually quote the maximum possible Z_e value in accordance with the Electricity, Safety, Quality and Continuity Regulations 2002; you will require permission to use this system.

3 Calculation: this is the most complex method and can only be used if the designer knows all of the relevant impedances and resistances of the system to be tested.

Three-phase supply for a TN-C-S system

The three-phase supply for a TN-C-S installation is shown in the diagram.

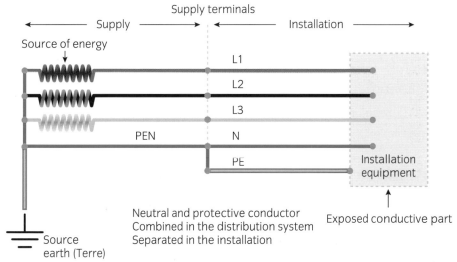

Diagram of a three-phase supply for a TN-C-S installation

Single-phase TN-C-S system

The diagram on page 78 is typical of a single-phase TN-C-S system for a domestic (or similar) installation.

In both three and single phase cases the electrical distributor will supply the consumer with an earth, which is connected to the neutral at the origin of the installation. The neutral is connected to the centre point of the supply transformer. As the neutral now doubles as an earth, it is referred to as a protective earthed neutral (PEN) conductor.

In the UK the most common earthing arrangement is now TN-C-S, or PME.

TN-C-S

T = source earth (*terre*).

N = neutral and protective conductor.

C = combined in the distribution system.

S = separated in the installation.

PME

Protective multiple earthing.

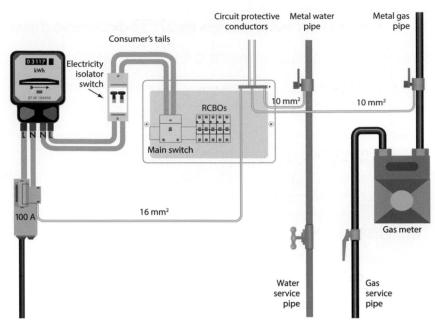

Practical arrangement for a single-phase TN-C-S system

New supplies will almost always be PME (protective multiple earthing) to provide for a TN-C-S system.

An enquiry to an electricity distributor for the characteristics of a domestic supply will result in the following information being provided:

- external earth fault loop impedance (Z_e) = 0.35 Ω, the maximum for PME supplies (TN-C-S systems)
- the Z_e value in practice will normally be considerably less than 0.35 Ω.

Protective multiple earthing (PME)

Protective multiple earthing (PME) is a system where multiple earth electrodes are installed to provide an alternative path to the star point of a sub-station transformer. As the PEN conductor is both a live conductor and earth, if this conductor becomes part of an open circuit in the supply, earthed parts of the installation may become live as current tries to find a path back to the sub-station. Remember, if current flows into a circuit, the same current flows back on the neutral. If this neutral becomes open, that current needs an alternative path and, as the earthing of the installation is connected to the supplier's neutral, this may provide the path needed.

If the supply contains multiple earth electrodes, this risk is reduced as the electrodes should provide a more suitable return path.

KEY POINT

Three methods of determining the external earth fault loop impedance are by measurement, enquiry or calculation.

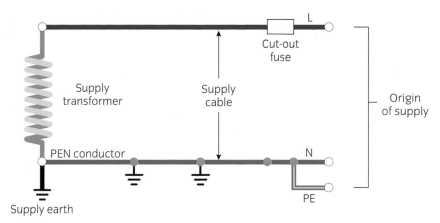

Simple representation of a TN-C-S system

ACTIVITY

Redraw the simple TN-C-S system diagram and illustrate how you would attach the probes or leads of a two-lead test meter to the supply to test the earth fault loop impedance (Z_e). Place as much information on the diagram as possible and show the maximum expected reading on the meter.

TT system and earth fault loop impedance

This section concentrates on the **TT** system.

There are three methods of determining the earth fault loop impedance (Z_e).

1 Measurement: by testing at the origin of the installation – this is the most common and accurate method used.

2 Enquiry: such as by contacting the electricity distributor, who will usually quote the maximum possible Z_e value in accordance with the Electricity, Safety, Quality and Continuity Regulations 2002.

3 Calculation: this is the most complex method and can only be used if the designer knows all of the relevant impedances and resistances of the system to be tested.

Three-phase supply for a TT system

The three-phase supply for a TT installation is shown in the diagram.

TT

T = source earth (*terre*).

T = installation earth (*terre*).

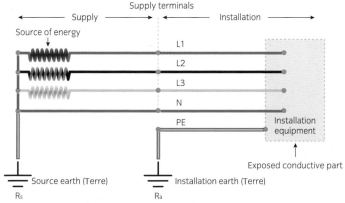

Diagram of a three-phase supply for a TT installation

A TT system is a system in which the electrical distributor does *not* supply the consumer with an earth. This may be because they can't guarantee the continuity of a protective conductor, such as in the case of a low voltage supply to a building by means of an overhead line which could deteriorate, break or become disconnected during adverse weather conditions.

TT systems may also be used on construction sites, caravan parks, marinas and many other locations. This is because the Electricity Safety Quality and Continuity Regulations will not permit the connection of PME supplies to certain locations. So, the supplier does not give an earth in these situations and it is up to the consumer to provide an earth using an earth electrode.

Generators may also use TT-type systems in order to provide an alternative, reliable earth path. If a generator is installed for standby purposes, and the power is lost, the generator starts up. This generator must have an independent earth as the supplier may be working on the supply system.

Single-phase supply for a TT system

The diagram shows a practical arrangement for a single-phase TT system. It is typical of a domestic (or similar) installation.

An enquiry to an electricity distributor for the supply characteristics of a domestic supply will result in the following information being provided:

■ external earth fault loop impedance (Z_e) = 21 Ω, the maximum value of the distributor's earth electrode and associated impedances of the supply transformer and line conductor (TT systems)

■ the value in practice will normally be considerably less than 21 Ω.

It is very important to remember that this value *does not include* the resistance of the consumer's earth electrode.

KEY POINT

Three methods of determining the external earth fault loop impedance are by measurement, enquiry or calculation.

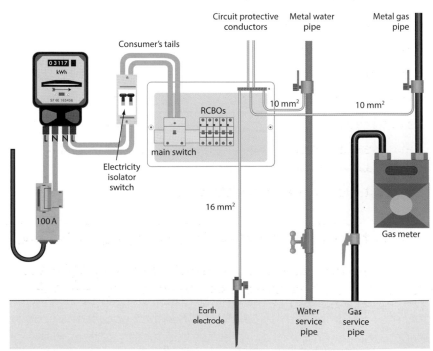

Practical arrangement for a single-phase TT system

THE CITY & GUILDS TEXTBOOK

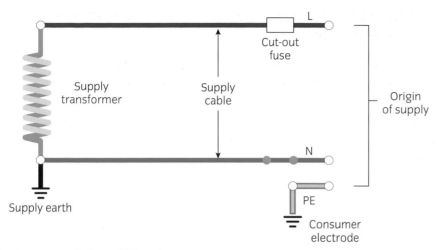

Simple representation of TT system

ACTIVITY

Redraw the simple TT system diagram and illustrate how you would attach the probes or leads of a two-lead test meter to the supply to test the earth fault loop impedance (Z_e). Place as much information on the diagram as possible and show the maximum expected reading on the meter.

EARTH ELECTRODE RESISTANCE TESTING

Earth electrode resistance testing can be carried out by using one of three test methods. This book will look at the two most common methods, using:

- a proprietary earth electrode resistance tester
- an earth fault loop impedance tester.

We will, for this book, refer to these methods as E1 and E2.

Assessment criteria

4.12 Specify and apply the methods for measuring earth electrode resistance and correctly interpreting the results

ASSESSMENT GUIDANCE

Most electricians are unlikely to carry out this test but it is important to know the principles of the test as you never know when you might need it.

Recognised types of earth electrode

The following types of earth electrode are recognised:

1 earth rods or pipes

2 earth tapes or wires

3 earth plates

4 underground structural metalwork, embedded in foundations

5 welded metal reinforcement of concrete, embedded in the earth

6 metal sheaths and coverings of cables (subject to Regulation 542.02 to 05 in BS 7671)

7 other suitable underground metalwork.

Reason for earth electrode resistance testing

The purpose of this test is to establish that the resistance of the soil surrounding an earth electrode is suitable and that the electrode makes contact with the soil.

Acceptable test values for an earth electrode

Earth electrode resistance values can differ greatly, dependent on the type of ground and environmental conditions, the material of the electrode used and area of contact with the general mass of earth.

It is recommended that the earth electrode resistance test is carried out when the ground conditions are least favourable, such as during dry weather.

Note that earth electrode resistance values above 200 Ω may not be stable, as soil conditions change due to factors such as soil drying and freezing.

Test method E1

This method uses a proprietary earth electrode resistance tester.

The test meter

Proprietary earth electrode resistance tester (image courtesy of Megger)

The photograph shows a typical four-terminal digital earth electrode resistance tester. Analogue (moving coil) meters can be used.

Note that this meter can only be used if the electrode is not yet connected to the installation or if the installation relying on this earth electrode is completely isolated and the earthing conductor is disconnected from the main earthing terminal.

Test connections C1 and P1 must be connected to the electrode under test.

Carrying out test E1

There are two possible methods used for test E1:

- Method 1 can be done by linking out C1 and P1 at the meter and supplying one lead only to the electrode, as shown in the diagram opposite.
- Method 2 is an alternative in which you can run two leads from the test meter to the electrode under test.

Method 2 may be used when ascertaining very low electrode values to eliminate the resistance of the test leads.

In most cases, method 1 will suffice as the lead resistance is generally less than 1 Ω. Always read the instrument manufacturer's instructions before use, if the instrument is not familiar to you.

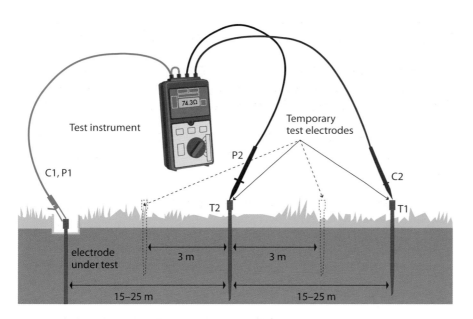

Typical earth electrode test using a three-or-four terminal tester (E1) (not to scale)

With the meter connected for method 1, there are two temporary test electrodes/spikes (T1 and T2) which must be inserted into the ground. These are normally supplied with the test instrument.

- C2 terminal on the meter is connected to T1 via a long lead, ideally 30–50 m away from the electrode under test.

- P2 terminal on the meter is connected to T2 via a long lead, and is centrally positioned between T1 and the electrode under test

Ideally, the distance between the earth electrode and test spike T1 should be ten times the length of the electrode under test, but this dimension is likely to be affected by the location of the electrode and any surrounding buildings, paths or driveways, for example.

Tests results

Three readings are taken during this test, with test spike T2 moved for each reading. The distance T2 is moved for the second and third readings depends on the distance between the electrode and spike T1. If the distance between them is 30 m then, typically, T2 will be moved 10% of that distance, which is 3 m. So, the first test is taken with spike T2 in the central position, the second test with the spike moved 10% closer to the earth electrode and the third test with the spike moved 10% from the centre, in the opposite direction to the earth electrode.

Here we will consider example readings for an earth electrode in good soil or clay (the earth electrode is 3 m long, so the distance between the electrode and test spike T2 is 30 m):

- with T2 central = 72 Ω

- with T2 3 m closer to the electrode under test = 70.5 Ω

- with T2 3 m closer to T1 = 73.5 Ω.

> **KEY POINT**
>
> Earth electrode resistance values can differ greatly, dependent on the type of ground and environmental conditions, the material of the electrode used and the area of contact with the general mass of earth.

Evaluation of test results

Once the three results have been obtained, the average of the three is found. So, using the example values given above, the average reading is:

$$\frac{72 + 70.5 + 73.5}{3} = 72\,\Omega$$

The three readings obtained should fall within a tolerance of 5% of the average, so 5% of 72 is 3.6 Ω so a tolerance of $\pm$ 5% gives 75.6 Ω and 68.4 Ω.

As the three readings all fall within this 5% tolerance, they are acceptable and the average value (72 Ω) would be recorded as the earth electrode resistance (R_A). If the deviation exceeds 5%, further tests must be carried out with a larger separation between the earth electrode under test and spike T1.

Test method E2

This method uses an earth fault impedance tester and can only be used where the installation has an available supply.

The test meter

The earth fault loop impedance tester and leads must comply with Guidance Note GS38.

Test results collected using this method will not be as accurate as those from a dedicated earth electrode tester.

Carrying out test E2

- The main switch for the installation must be securely isolated.
- Disconnect the earthing conductor from the main earthing terminal (MET) or the earth bar, if no MET is present. At this point, for safety reasons, check all main protective bonding within the installation is securely connected together at the MET.
- Ensure that there are no parallel earth paths for the test current. The earthing conductor must connect directly to the earth electrode under test and no other parts.
- Set the meter for earth loop impedance (Z_e on some meters). Check that the leads are in good condition and applicable for the tests.
- Attach the earth clip to the disconnected earthing conductor which connects to the electrode under test.
- Locate the probe to the incoming line conductor terminal on the supply side of the main switch.
- Press the test button and record the reading.

Note that some test instruments require connection of a third test probe to the neutral of the supply.

Repeat this test to ensure an accurate reading has been achieved. This reading can be accepted as your earth electrode resistance. It is recommended that this reading should be below 200 Ω to allow for soil drying and freezing.

Test results

When using the above test method, the result will be a combination of the consumer's earth electrode resistance and the associated resistances of the line conductor, the impedance of the supply transformer, the supplier's earth electrode and the soil between the supplier's and consumer's electrodes. As this test method measures the overall loop impedance as described, we will refer to it as Z_s.

The diagram shows the method of testing the earth electrode using an earth fault loop impedance tester (method E2).

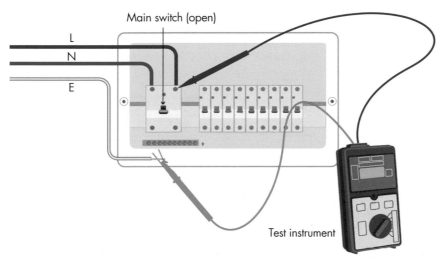

Earth fault loop impedance tester (E2)

Verification of test results

For the purpose of verification of compliance, assume that a test result for an earth electrode when testing using an earth fault loop impedance tester was 115 Ω (Z_s).

Regulation 411.5.3 requires:

$$R_A \times I_{\Delta n} \leq 50 \text{ V}$$

or

$$Z_s \times I_{\Delta n} \leq 50 \text{ V}$$

where R_A = the sum of electrode and protective conductor to exposed conductive parts

Z_s = the earth fault loop impedance

$I_{\Delta n}$ = RCD rated residual operating current

Table 41.5 from BS 7671 helps us with the answers to such calculations, as long as the value recorded for Z_s is below that given in the table for the particular residual current setting ($I_{\Delta n}$) of the RCD protecting the installation. (Although the table only states Z_s, and not R_A, as the formulae above are similar, either value could be used.)

If we assume, for each of our test methods that a 100 mA RCD protected each installation, we can see from the table that both results are acceptable since they are both less than 500 Ω.

RCD rated residual operating current (mA)	Maximum value of earth fault loop impedance, Z_s (Ω)
30	1667
100	500
300	167
500	100

Table 41.5 Maximum value of earth fault loop impedance for non-delayed RCDs

Equally, using this table you can see that you could use a 30 mA, 100 mA or 300 mA RCD as a main isolation switch – so the test results could be used to determine a suitably rated RCD to protect the installation.

The Z_s value recorded for method E2 is too high to install a 500 mA RCD, but this RCD could be used where we recorded 72 Ω as a value of R_A.

Discrimination between RCDs

Due to the design of the final circuits and their associated protection devices, if you select a 300 mA RCD as a main switch, you must ensure that there is discrimination between RCDs. This can be achieved by using a time delay S-type RCD.

If S-type RCDs are used, careful consideration must be given to ensure that requirements of Regulation 411.3 and 411.5 are met. Manufacturers' data must be verified for time delay criteria.

KEY POINT

The installation must be isolated and the earthing conductor must be disconnected from the MET to avoid parallel earth paths during this test, whichever method is used.

ASSESSMENT GUIDANCE

As part of assessment, you may be asked to determine the maximum Z_s allowed for an RCD.

ACTIVITY

Calculate the maximum Z_s allowed for the following RCDs, without using Table 41.5. The first one has been completed as an example.

- 30 mA RCD
  $$\frac{50 \text{ V}}{0.03 \text{ A}} = 1667 \,\Omega$$
- 100 mA RCD
- 300 mA RCD
- 500 mA RCD

AUTOMATIC DISCONNECTION OF SUPPLY

Assessment criteria

4.14 State the methods for verifying protection by automatic disconnection of the supply

TN systems

A TN system, in this case, can be either a TN-S or TN-C-S system. In a TN system, the electrical distributor guarantees a good earth at the origin of the supply.

It is important to note that, if an earth fault of negligible impedance develops within the fixed wiring or equipment of an installation, the protective device used to protect that circuit must operate within certain time limits given in Chapter 41 of BS 7671. This is called protection by automatic disconnection of supply.

Earth fault loop impedance circuit design and calculation for TN systems

The earth fault loop impedance in part of a TN-C-S system is shown in the diagram.

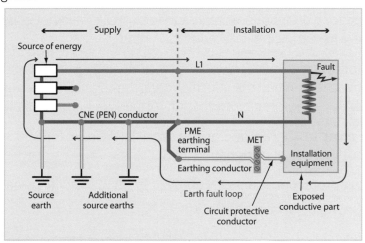

Earth fault loop impedance path of a TN-C-S system

Protection of TN systems by automatic disconnection of supply

As part of the design process of all circuits, evaluation of the earthing and associated cables must be considered to ensure safe disconnection during earth fault conditions.

The effectiveness of measures for fault protection by automatic disconnection of supply can be verified for installations within a TN system by:

- measurement of earth fault loop impedance
- confirmation by visual inspection that overcurrent devices use suitable short-time or instantaneous tripping settings for circuit-breakers, correct current rating (I_n) and type for fuses and correct coordination and settings for RCDs
- testing to confirm that the disconnection times set out in Chapter 41 of BS 7671 can be achieved.

Maximum disconnection times for a TN system

Table 41.1 from Chapter 41 of BS 7671 (shown below) gives the maximum disconnection times for final circuits not exceeding 32 A.

From the table, we can see that the maximum disconnection time for a final circuit forming part of a TN system where the nominal voltage to earth (V_0) is 230 V, is 0.4 seconds.

TABLE 41.1
Maximum disconnection times

System	50 V < V_0 ≤ 120 V seconds		120 V < V_0 ≤ 230 V seconds		230 V < V_0 ≤ 400 V seconds		V_0 > 400 V seconds	
	a.c.	d.c.	a.c.	d.c.	a.c.	d.c.	a.c.	d.c.
TN	0.8	NOTE	0.4	5	0.2	0.4	0.1	0.1
TT	0.3	NOTE	0.2	0.4	0.07	0.2	0.04	0.1

Maximum disconnection times for TN and TT systems

In a TN system, a disconnection time not exceeding 5 seconds is permitted for a distribution circuit and for a final circuit not covered by the table where the circuit exceeds 32 A rating.

For the following calculation, look again at the location plan of the small industrial unit and the information given regarding the supply characteristics and circuit information (Appendix 1 and Appendix 2).

You are going to concentrate on the radial final circuit for BS 1363 socket outlets, circuit designation 'Bl-3'.

The circuit protection is a BS EN 60898, type-B 32 A circuit breaker.

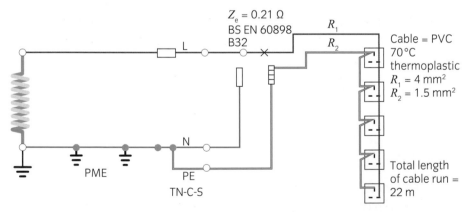

Five radial socket outlets in a circuit

The maximum disconnection time allowed for this circuit is 0.4 seconds, because the circuit is a final circuit and it does not exceed 32 A (see Table 41.1 above).

Impedance values by design and calculation

With reference to the diagram on page 88, when you have established the cable length and cross-sectional area, you can begin to calculate the resistance of the line conductor (R_1) and the cpc (R_2).

You need to use the formula:

$$Z_s = Z_e + (R_1 + R_2)$$

$Z_e = 0.21\ \Omega$ (given value at the DB)

Using a BS EN 60898, type-B 32 A circuit breaker the maximum Z_s allowed = 1.44 Ω (taken from Table 41.3 shown below)

TABLE 41.3

(a) Type B circuit-breakers to BS EN 60898 and the overcurrent characteristics of RCBOs to BS EN 61009-1

Rating (amperes)	3	6	10	16	20	25	32	40	50	63	80	100	125	I_n
Z_s (ohms)		7.67		2.87		1.84		1.15		0.73		0.46		$46/I_n$
	15.33		4.60		2.30		1.44		0.92		0.57		0.37	

(b) Type C circuit-breakers to BS EN 60898 and the overcurrent characteristics of RCBOs to BS EN 61009-1

Rating (amperes)	6	10	16	20	25	32	40	50	63	80	100	125	I_n
Z_s (ohms)	3.83		1.44		0.92		0.57		0.36		0.23		$23/I_n$
		2.30		1.15		0.72		0.46		0.29		0.18	

(c) Type D circuit-breakers to BS EN 60898 and the overcurrent characteristics of RCBOs to BS EN 61009-1

Rating (amperes)	6	10	16	20	25	32	40	50	63	80	100	125	I_n
Z_s (ohms)	1.92		0.72		0.46		0.29		0.18		0.11		$11.5/I_n$
		1.15		0.57		0.36		0.23		0.14		0.09	

Maximum earth fault loop impedence (Z_s) for circuit breakers to comply with 0.4 s and 5 s disconnection times (from BS 7671:2008(2011))

ASSESSMENT GUIDANCE

Don't forget that the figures for earth loop impedance are maximums and to compare them to measured values they should be multiplied by the rule of thumb figure of 0.8 (80%).

Establishing the resistance of $R_1 + R_2$

You need to establish the $R_1 + R_2$ of this cable by using the table below, which is a small section of Table B1 in Guidance Note 3. The cable installed for this circuit is a 4 mm²/1.5 mm². Once the $R_1 + R_2$ is calculated, you can add this value to the Z_e value. We will assume an ambient temperature of 20 °C.

▼ **Table B1** Values of resistance/metre for copper and aluminium conductors and of ($R_1 + R_2$) per metre at 20 °C in milliohms/metre

Cross-sectional area (mm²)		Resistance/metre or ($R_1 + R_2$)/metre (mΩ/m)	
Line conductor	Protective conductor	Copper	Aluminium
1	–	18.10	
1	1	36.20	
1.5	–	12.10	
1.5	1	30.20	
1.5	1.5	24.20	
2.5	–	7.41	
2.5	1	25.51	
2.5	1.5	19.51	
2.5	2.5	14.82	
4	–	4.61	
4	1.5	16.71	
4	2.5	12.02	
4	4	9.22	

Resistances of copper conductors (from Guidance Note 3: Inspection and Testing, IET)

$R_1 + R_2$ per metre $= 16.71$ mΩ

Total $R_1 + R_2 = 22$ m of 4 mm^2/1.5 mm^2 cable is determined by

$$\frac{22 \times 16.71}{1000} = 0.37 \ \Omega \text{ at } 20\,°\text{C}$$

If the cable is carrying its full load current, the temperature of the cable could increase to 70 °C. An increase in temperature will increase the cable's resistance.

By applying the applicable multiplier of 1.2, from the table below, we can see that the $R_1 + R_2$ value would be a maximum of

0.37 Ω x 1.20 = 0.44 Ω at normal operating temperatures.

The 1.2 factor was selected as the cables are 70 °C thermoplastic and the cpc is bunched or incorporated with the live conductors.

▼ **Table B3** Conductor temperature factor F for standard devices

Multipliers to be applied to Table B1 for devices in Tables 41.2, 41.3, 41.4

Conductor installation	Conductor insulation		
	70°C thermoplastic (PVC)	85°C thermosetting (note 4)	90°C thermosetting (note 4)
Not incorporated in a cable and not bunched (notes 1, 3)	1.04	1.04	1.04
Incorporated in a cable or bunched (notes 2, 3)	1.20	1.26	1.28

Conductor temperature factor F for standard devices – multipliers to be applied to Table B1 for devices in BS 7671 Tables 41.2, 41.3 and 41.4 (Tables B1 and B3 are from Guidance Note 3: Inspection and Testing, IET)

$Z_s = Z_e + (R_1 + R_2)$

$\quad = 0.21 \ \Omega + 0.44 \ \Omega$

$\quad = 0.65 \ \Omega$ which is less than the maximum Z_s permitted by Table 41.3.

(Remember, the maximum Z_s permitted $= 1.44 \ \Omega$)

Verification of earth fault loop impedance test results

BS 7671 requires the inspector not only to test the installation, but also *to compare* the results with relevant design criteria (or with criteria within BS 7671). This may seem obvious, but some inspectors do pass test information back to their office without making the necessary comparisons, possibly assuming that someone else will check the results.

Disconnection times for a TN system

Table 41.1 from Chapter 41 of BS 7671 gives the maximum disconnection times for final circuits not exceeding 32 A. From the table (opposite), we can see that the maximum disconnection time for a final circuit forming part of a TN system where the nominal voltage to earth (V_0) is 230 V, is 0.4 seconds.

ACTIVITY

Using this method of design, calculate the Z_s for circuit 'Gr-3', instantaneous water heater (see Appendix 1 and Appendix 2).

Verify that the Z_s is within the maximum value permitted in Table 41.3.

The practical testing of earth fault loop impedance for this circuit is shown next.

In a TN system, a disconnection time not exceeding 5 seconds is permitted for a distribution circuit and for a final circuit not covered by the table.

TABLE 41.1
Maximum disconnection times

System	$50 V < V_0 \leq 120 V$ seconds		$120 V < V_0 \leq 230 V$ seconds		$230 V < V_0 \leq 400 V$ seconds		$V_0 > 400 V$ seconds	
	a.c.	d.c.	a.c.	d.c.	a.c.	d.c.	a.c.	d.c.
TN	0.8	NOTE	0.4	5	0.2	0.4	0.1	0.1
TT	0.3	NOTE	0.2	0.4	0.07	0.2	0.04	0.1

Maximum disconnection times for TN and TT systems (from BS 7671:2008 (2011))

For the following calculation, look again at the location plan of the small industrial unit and the information given regarding the supply characteristics and circuit information, in Appendix 1.

You are going to concentrate on the radial final circuit for BS 1363 socket outlets, circuit designation (Bl-3). The circuit protection is a BS EN 60898, type-B 32 A circuit breaker.

The maximum disconnection time allowed for this circuit is 0.4 seconds because the circuit is a final circuit and it does not exceed 32 A (see Table 41.1).

Methods of measuring earth fault loop impedance

Earth fault loop impedance values for final circuits can be obtained in two ways

- adding Z_e to $R_1 + R_2$
- direct measurement.

In both cases, the value of Z_e at the origin of an installation needs to be established and the most accurate method is by direct measurement. We have seen how we can obtain values of Z_e for TT installations on page 79.

Practical guide to testing the Z_e at the origin of a TN system

The diagram on page 92 provides a simple representation of a TN-S system that highlights the external circuit from the supply transformer to the origin of the consumer's supply.

The external earth fault loop impedance (Z_e) in ohms is a result of the impedance of the supplier's line conductor, the transformer and the separate earth added together.

> **ASSESSMENT GUIDANCE**
>
> Remember that the earth loop test is a live test and you must take all precautions necessary.

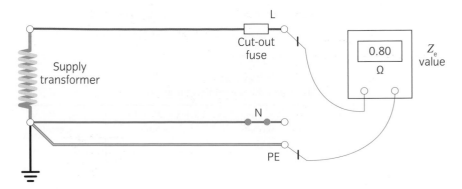

The Z_e value illustrated is the maximum value of the earth fault loop impedance expected for a TN-S system.

Simple diagrammatic representation of a TN-S system

The test is carried out in the following way using an earth fault loop impedance tester and leads which must comply with Guidance Note GS38.

- Isolate and secure the installation main switch in the off position.
- Ensure, for reasons of safety, that all main protective bonding is connected to the main earthing terminal. If the supplier's cable is faulty, a fault current may be introduced to extraneous parts of the installation.
- Disconnect the earthing conductor from the main earthing terminal.
- Check instrument for safety and correct settings.
- Connect the earth clip to the disconnected earthing conductor, then connect the test instrument line probe to the supply line terminal of the main switch.
- Press the test button and note the result.
- For three-phase installations, repeat this test for all line conductors. For three-phase installations, the highest reading obtained is recorded as the external earth fault loop impedance.
- For a single-phase installation, record the value.
- Reconnect the earthing conductor to the MET.
- The installation supply may be switched back on if required.

Note that some test instruments may require a third lead connected to the neutral of the supply.

KEY POINT

Three methods of determining the external earth fault loop impedance are by measurement, enquiry or calculation.

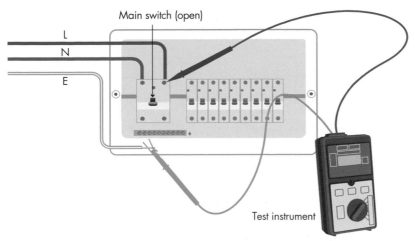

Example test of Z_e at the origin of an installation

Determining values of Z_s by calculation

As we have now determined the value of Z_e at the origin of the installation, we can determine the total earth fault loop impedance (Z_s) by adding the R_1+R_2 which we previously obtained through continuity testing.

SmartScreen Unit 307
Handout 34 and Worksheet 34

Providing continuity testing is carried out thoroughly, which it should have been for any initial verification, this method will provide reliable values of Z_s, even if parallel earth paths change due to work carried out on mechanical services. For this reason, and for reasons of testing safely, this is the preferred method of obtaining values of Z_s. This method also reduces the need to revisit parts of the installation which have already been tested.

If values of Z_s are obtained through direct measurement (main protective bonding must remain connected during the test, as these provide parallel earth paths), the values obtained by this method may not be a true reflection of the circuit conditions. Sample testing by direct measurement is a good exercise once Z_s values are obtained by calculation; the values obtained by direct measurement are likely to be lower than those calculated.

Method for obtaining Z_s by direct measurement

1 Select the correct meter – you will need an earth fault loop impedance tester.

2 Select the correct scale – the correct scale is loop, Z_e or Z_s scale (depending on the instrument used).

3 Check the leads for damage – ensure that the probe tips are protruding no more than 4 mm (preferably 2 mm) in accordance with Guidance Note GS38.

4 Insert leads – choose the correct location on the meter (following the manufacturer's instructions).

5 Connect the leads correctly.

You can choose either the appropriate plug-in lead for testing socket outlets or fly leads appropriate to the terminals which need to be tested. Remember that all test leads must be manufactured and maintained in accordance with Guidance Note GS38.

You are now ready to carry out the earth fault loop impedance testing.

Earth fault loop impedance (Z_s) testing for radial socket outlets

Each socket outlet must be tested, with the highest value of Z_s recorded on the Generic Schedule of Test Results.

The meter in the diagram is shown at the final socket outlet on the radial circuit. This will give the Z_s value if all of the conductor connections are secure.

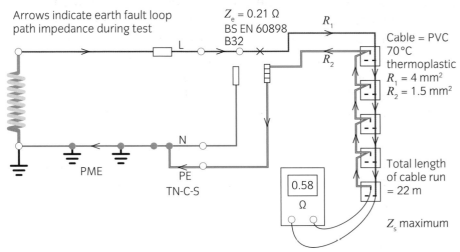

Five radial socket outlets with 4 mm/1.5 mm cables

Verification of earth fault loop impedance (Z_s) for TN systems

Once you have established the Z_s, you can compare this value to the maximum value allowed, to ensure the disconnection time can be achieved.

The Z_s in the circuit above = 0.58 Ω.

The maximum earth fault loop impedance allowed for this circuit can be found by using the table opposite. The table can be used to find the maximum Z_s for disconnection times of 0.1 to 5 seconds for TN systems.

This table can be used for the following overcurrent protective devices: BS 3871, BS EN 60898 circuit breakers and BS EN 61009 RCBOs (residual current operated circuit breakers with integral overcurrent protection).

Circuit-breaker type	Circuit-breaker rating (A)													
	5	6	10	15	16	20	25	30	32	40	45	50	63	100	125
1	9.27	7.73	4.64	3.09	2.90	2.32	1.85	1.55	1.45	1.16	1.03	0.93	0.74	0.46	0.37
2	5.3	4.42	2.65	1.77	1.66	1.32	1.06	0.88	0.83	0.66	0.59	0.53	0.42	0.26	0.21
B	7.42	6.18	3.71	2.47	2.32	1.85	1.48	1.24	1.16	0.93	0.82	0.74	0.59	0.37	0.30
3&C	3.71	3.09	1.85	1.24	1.16	0.93	0.74	0.62	0.58	0.46	0.41	0.37	0.29	0.19	0.15
D	1.85	1.55	0.93	0.62	0.58	0.46	0.37	0.31	0.29	0.23	0.21	0.19	0.15	0.09	0.07

Maximum Z_s test values in ohms for circuits supplied by circuit breakers (from Guidance Note 3: Inspection and Testing, IET)

The maximum Z_s allowed for a type-B 32 A circuit breaker is 1.16 Ω.

The maximum Z_s test result = 0.58 Ω.

We can conclude that the tested Z_s is well within the 0.4 seconds allowed.

Note this table specifies that the ambient temperature during the testing should be no lower than 10 °C and is, therefore, temperature corrected. No further calculation is required if the Z_s test result is lower than the maximum allowed in this table, unless the ambient temperature is lower than 10 °C. It is advisable to log the test temperature, as this allows future test results to be evaluated accurately. This table is also shown in the IET On-Site Guide.

If the maximum values of Z_s are used, as given in BS 7671, correction for temperature will be required. As the values in BS 7671 are used for design purposes, not verification, the values must be corrected (reduced) by 20% to allow for temperature changes under operating conditions.

As a rule of thumb, the measured values of Z_s should not exceed 80% of the values given in Tables 41.2 to 41.4 in BS 7671.

As an example, the maximum permitted value of earth fault loop impedance for a type-B 32 A circuit breaker is 1.44 Ω; 80% of this is (1.44 × 0.8) = 1.16 Ω. Any measured value should not exceed this.

So measured Z_s should be ≤ maximum Z_s as BS 7671 × 0.8

If you get a higher Z_s test reading than the maximum allowed, the circuit $R_1 + R_2$ must be reduced or the circuit components or the protective device may need to be redesigned or changed:

- By increasing the cross-sectional area of the cable you will reduce the $R_1 + R_2$ and therefore reduce the total Z_s value.

- By changing the circuit breaker to an RCBO or by using an RCD to supply the circuit in accordance with Regulation 411.4.9, you can achieve disconnection within the 0.4 seconds.

KEY POINT

It is very important to understand the theory of the maximum Z_s values for circuits. You must ensure that, under fault conditions, the protective device will disconnect within the maximum duration limits for safety.

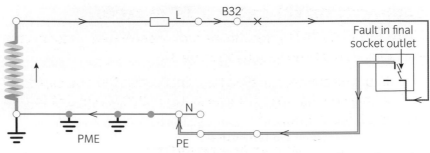

Basic diagram of the earth fault loop impedance path at a final socket outlet to the transformer

This diagram illustrates the earth fault loop path of the radial socket outlet circuit. You can use this method to illustrate the earth fault loop impedance of fixed equipment or lighting circuits.

TT systems

Verification of earth fault loop impedance test results in TT system

It is important to note that, if an earth fault of negligible impedance develops within the fixed wiring or equipment of an installation, the protective device must operate within the time limits given in Chapter 41 of BS 7671. This is called protection by automatic disconnection of supply.

BS 7671 requires the inspector not only to test the installation, but also *to compare* the results with relevant design criteria (or with criteria within BS 7671). This may seem obvious, but some inspectors do pass test information back to their office without making the necessary comparisons, possibly assuming that someone else will check the results.

Protection of TT systems by automatic disconnection of supply

As part of the design process of all circuits, evaluation of the earthing and associated cables must be considered to ensure safe disconnection during earth fault conditions.

The effectiveness of measures for fault protection by automatic disconnection of supply can be verified for installations within a TT system by:

- measurement of earth fault loop impedance
- confirmation by visual inspection that overcurrent devices use suitable short-time or instantaneous tripping setting for circuit-breakers, correct current rating (I_n) and type for fuses and correct coordination and settings for RCDs
- testing to confirm that the disconnection times set out in Chapter 41 of BS 7671 can be achieved.

Disconnection times for a TT system

Table 41.1 from BS 7671 gives the maximum permitted disconnection times for final circuits up to and including 32 A.

The maximum disconnection time for a final circuit forming part of a TT system, with a nominal voltage to earth (U_0) of 230 V with a rating not exceeding 32 A is 0.2 seconds.

TABLE 41.1
Maximum disconnection times

System	50 V < V_0 ≤ 120 V seconds		120 V < V_0 ≤ 230 V seconds		230 V < V_0 ≤ 400 V seconds		V_0 > 400 V seconds	
	a.c.	d.c.	a.c.	d.c.	a.c.	d.c.	a.c.	d.c.
TN	0.8	NOTE	0.4	5	0.2	0.4	0.1	0.1
TT	0.3	NOTE	0.2	0.4	0.07	0.2	0.04	0.1

Maximum disconnection times for TN and TT Systems (from BS 7671:2008 (2011))

In a TT system, a disconnection time not exceeding 1 second is permitted for a distribution circuit and for a final circuit not covered by Table 41.1.

Verification of test results

BS 7671 requirements for TT installations are fairly straightforward.

Regulation 411.5.3 requires:

$$R_A \times I_{\Delta n} \leq 50 \text{ V}$$
$$\text{or}$$
$$Z_s \times I_{\Delta n} \leq 50 \text{ V}$$

For the purpose of verification of compliance, assume that the test result for earth fault loop impedance including the earth electrode resistance was 113 Ω (Z_e).

If the lighting circuit has an $R_1 + R_2$ value of 2 Ω, we can assume the total Z_s to be:

$$Z_s = Z_e + R_1 + R_2$$
$$= 113 \, \Omega + 2 \, \Omega$$
$$= 115 \, \Omega$$

Regulation 411.5.3 requires:

$$R_A \times I_{\Delta n} \leq 50 \text{ V}$$
$$\text{or}$$
$$Z_s \times I_{\Delta n} \leq 50 \text{ V}$$

R_A = the sum of electrode and protective conductor to exposed conductive parts (electrode resistance added to circuit $R_1 + R_2$)

Z_s = the earth fault loop impedance

$I_{\Delta n}$ = RCD rated residual operating current

Maximum R_A values can be substituted for Z_s in this case.

Non-delayed RCD rated residual operating current $I_{\Delta n}$ (mA)	Maximum value of earth fault loop impedance, Z_s (Ω)
30	1667
100	500
300	167
500	100

Maximum R_A and Z_s values for non-delayed residual current devices

This table shows that you could use a 30 mA, 100 mA or 300 mA RCD as a protective device for this circuit. Note that the Z_s value is too high to install a 500 mA RCD. You would normally select a 30 mA RCBO for this circuit due to requirements for additional protection.

The guaranteed design fault current will be:

$$I_{fault} = \frac{V_0}{Z_s} = \frac{230\ V}{115\ \Omega} = 2\ A$$

Fault current in relation to the disconnection time of an RCD

The fault current is calculated to be 2 A or 2000 mA.

All non-delayed 30 mA RCDs have been manufactured to operate within 40 ms (0.04 seconds) at a 150 mA fault current. Therefore, disconnection for this circuit will occur well within the maximum 0.2 seconds disconnection time shown in Table 41.1.

Discrimination between RCDs

Due to the design of the final circuits and their associated protection devices, if you select a 300 mA RCD as a main switch, you must ensure that there is discrimination between RCDs. This can be achieved by using a time delay S-type RCD.

The diagram opposite shows a fault on an item of Class I (earthed) equipment.

- RCD A could be a main isolator for the distribution board (300 mA $I_{\Delta n}$ S-type).
- RCD B could be a 30 mA $I_{\Delta n}$ non-delayed RCBO for the lighting circuit.

If an earth fault develops on the luminaire (Class I equipment), RCD B will trip and the supply to the other circuits will remain healthy. If S-type RCDs are used, careful consideration must be given to ensure that requirements of Regulations 411.3 and 411.5 of BS 7671 are met. Manufacturers' data must be verified for time delay criteria.

KEY POINT

It is very important to understand the theory of the maximum Z_s values for circuits. You must ensure that under fault conditions, the protective device will disconnect within the maximum duration limits for safety.

ASSESSMENT GUIDANCE

You may be asked to calculate the maximum Z_s allowed for an RCD.

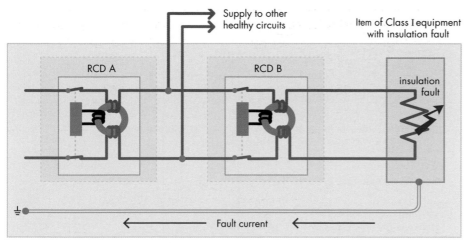

A fault on an item of Class I (earthed) equipment

PROSPECTIVE FAULT CURRENT (I_{PF})

The prospective short-circuit current (PSCC) and prospective earth fault current (PEFC) should be measured, calculated or determined by another method (such as by enquiry to the relevant electrical distributor), at the origin and at other relevant points in the installation.

Evaluating the prospective fault current

You need to evaluate the prospective fault current to ensure that the switchgear and protective devices are capable of withstanding high fault currents without damage.

You evaluate the prospective fault current at the origin and other relevant points of the installation, i.e. main distribution boards, sub-main distribution boards and consumer units.

Regulation 434.5.1 in BS 7671 states that the **breaking capacity** rating of each protective device shall be not less than the prospective fault current at its point of installation.

The table on page 100 gives an indication of the maximum breaking capacities of our most commonly used protection devices.

ASSESSMENT GUIDANCE

Remember this is the prospective *maximum* fault current. It is quite likely that the fuse will blow or the circuit breaker will trip before this figure is reached. You cannot, however, reduce the rating of the device.

Breaking capacity

The amount of current a protective device can safely disconnect.

Device type	Device designation	Rated short-circuit capacity (kA)	
Semi-enclosed fuse to BS 3036 with category of duty	S1A S2A S4A	1 2 4	
Cartridge fuse to BS 1361 type I type II		16.5 33.0	
General purpose fuse to BS 88-2		50 at 415 V	
BS 88-3 type I type II		16 31.5	
General purpose fuse to BS 88-6		16.5 at 240 V 80 at 415 V	
Circuit-breakers to BS 3871 (replaced by BS EN 60898)	M1 M1.5 M3 M4.5 M6 M9	1 1.5 3 4.5 6 9	
Circuit-breakers to BS EN 60898* and RCBOs to BS EN 61009		I_{cn} 1.5 3.0 6 10 15 20 25	I_{cs} (1.5) (3.0) (6.0) (7.5) (7.5) (10.0) (12.5)

* Two short-circuit capacities are defined in BS EN 60898 and BS EN 61009:

I_{cn} – the rated short-circuit capacity (marked on the device).

I_{cs} – the in-service short-circuit capacity.

Rated short-circuit capacities of protective devices (from On-Site Guide, IET)

The circuit breaker above shows a 3 kA rated BS EN 60898

When evaluating the kA value of BS EN 60898 circuit breakers, you need to look for the number in the rectangular box. If you see a symbol like this:

6000

it means that the circuit breaker has been manufactured to be able to withstand 6 kA. A domestic location fault current is unlikely to exceed 6 kA.

How to check the prospective fault current

You can check the prospective fault current by measuring it during the inspection and testing process, ensuring the recorded values of I_{pf} are lower than the breaking capacities of the devices at that point or further into the installation.

Setting up the test meter for testing the prospective fault current

Most earth fault loop impedance testers on the market are capable of evaluating the prospective fault current. Always follow the manufacturer's instructions.

Method

1 Select the correct meter – the correct meter is a prospective fault current tester or an earth fault loop impedance tester (this will require calculation).

2 Select the correct scale – the correct scale is PSC/kA or kA, loop, or the Z_e or Z_s scale (depending on the individual instrument). Always read the manufacturer's instructions.

3 Check the leads for damage – ensure that the probe tips are protruding no more than 4 mm (preferably 2 mm) and all other parts are in accordance with Guidance Note GS38.

4 Insert leads – choose the correct location on the meter for the connection of the correct leads (follow the manufacturer's instructions).

Carrying out the tests for prospective fault current (PFC)

Testing the prospective earth fault current (PEFC)

■ Ensure that the main protective bonding and earthing conductor are securely connected. It is important to ensure *all* possible parallel paths exist for this test, as this will produce the maximum fault current.

■ Connect the earth clip of the instrument to the main earthing terminal (MET). For many three-lead instruments, the neutral clip can also be connected to the MET.

■ Connect the line probe of the instrument to the incoming line connection.

■ Take a measurement and note this as the prospective earth fault current (PEFC).

■ For three-phase installations, repeat this test for all line conductors individually, and record the highest value.

■ Remove all connections.

Testing the prospective short circuit current (PSCC)

■ Connect the earth and neutral clips to the neutral bar or connection.

■ Connect the line probe of the instrument to the incoming line connection.

■ Take a measurement and note this as the prospective short circuit current (PSCC).

■ For three-phase installations, repeat this test for all line conductors individually, and record the highest value.

■ Remove all connections.

Although some test instruments can measure fault currents between phases, do not attempt this unless you are absolutely certain of the instrument's capabilities. Remember that all test leads must be manufactured and maintained in accordance with Guidance Note GS38.

Determining the prospective fault current

Once the readings have been taken, the value to be recorded as prospective fault current is determined in the following way:

- For single-phase installations – determine which value from PEFC and PSCC is the largest and record that value as the prospective fault current. If readings were obtained in ohms, divide 230 V by the lowest value (Ohm's law) to obtain the prospective fault current.

- For three-phase installations – as the voltage between phases is 400 V, but the measured value was obtained using 230 V, the PSCC has to be adjusted. It is acceptable to *double* the largest value of PSCC recorded between the line conductors and neutral. This will normally be larger than any PEFC, so would be recorded as the prospective fault current.

For example, when testing a three-phase installation for both PEFC and PSCC, the results shown in the table were obtained.

L1 to E	PEFC	1.2 kA	Largest value is 1.21 kA
L2 to E	PEFC	1.21 kA	
L3 to E	PEFC	1.19 kA	
L1 to N	PSCC	1.42 kA	Largest value is 1.47 kA
L2 to N	PSCC	1.47 kA	
L3 to N	PSCC	1.39 kA	

ASSESSMENT GUIDANCE

Virtually every device, except the lowest rated rewirable devices, will be able to handle the values shown in the table.

The value to be recorded is:

$$1.47 \text{ kA} \times 2 = 2.94 \text{ kA}$$

Further tests for the prospective fault current

If the value of prospective fault current obtained at the origin of the installation is lower than the breaking capacity of any device within the installation at any remote distribution board or consumer unit, no further testing is required.

If, however, the prospective fault current at the origin is greater than the breaking capacity of devices at a remote distribution board, further testing, as described above, would need to be carried out at that particular distribution board. The fault current at that board is likely to be reduced due to the resistance or impedance of any distribution circuit.

Remember, the purpose of testing for prospective fault current is to ensure that protective devices can safely respond to any fault current they may experience at that point in the installation.

Calculating the fault current value

The fault current value can be calculated during the design process. This method tends to be used on larger projects only.

A designer of a large installation will gather information about the external impedances of the supply transformer and associated conductors. Using this information, the designer will then evaluate the maximum prospective fault current value.

Determining the fault current value by enquiry

You can determine the fault current value from the electricity distributor, by enquiry (usually by telephone). The distributor will require the address of the property, allowing them to assess the plans of the wiring system that supplies the property and enabling them to estimate the maximum prospective fault current at the origin of the supply.

In the absence of any plans, for 230 V single-phase supplies up to 100 A, the distributor will provide the consumer with an estimate of the maximum prospective short-circuit current at the distributor's cut-out, which will be based on Engineering Recommendation P25/1 and on the declared level of 16 kA at the point of connection of the service line to the LV distribution cable.

The fault level will only be this high if the installation is close to the distribution transformer. However, because changes may be made to the distribution network by the distributor over the life of an installation, the designer of the consumer's installation must specify equipment suitable for the highest fault level likely.

Remember that 16 kA will be the maximum value quoted. In most cases the I_{pf} will be a lot less than this value.

OPERATION AND FUNCTIONAL TESTING OF RESIDUAL CURRENT DEVICES (RCDS)

RCDs must be tested to ensure that they are working in accordance with BS and BS EN manufacturing standards.

The theory of RCD operation

'Residual current device' or 'RCD' is the generic term for a device that operates when the residual current in the circuit reaches a predetermined value. An RCD is a protective device used to automatically disconnect the electrical supply when an imbalance is detected between the line and neutral conductors.

ACTIVITY

Test the prospective fault current at a distribution board and check that the protective devices are capable of withstanding the maximum fault current by carrying out a visual inspection. Ensure the supply is securely isolated before any inspection is carried out. This activity must be carried out under the supervision of your tutor or assessor.

ASSESSMENT GUIDANCE

As part of the practical assessment, you will be required to test and verify prospective fault current.

Assessment criteria

4.16 Specify the methods for testing the correct operation of residual current devices (RCDs)

The diagram shows a line to earth (cpc) fault.

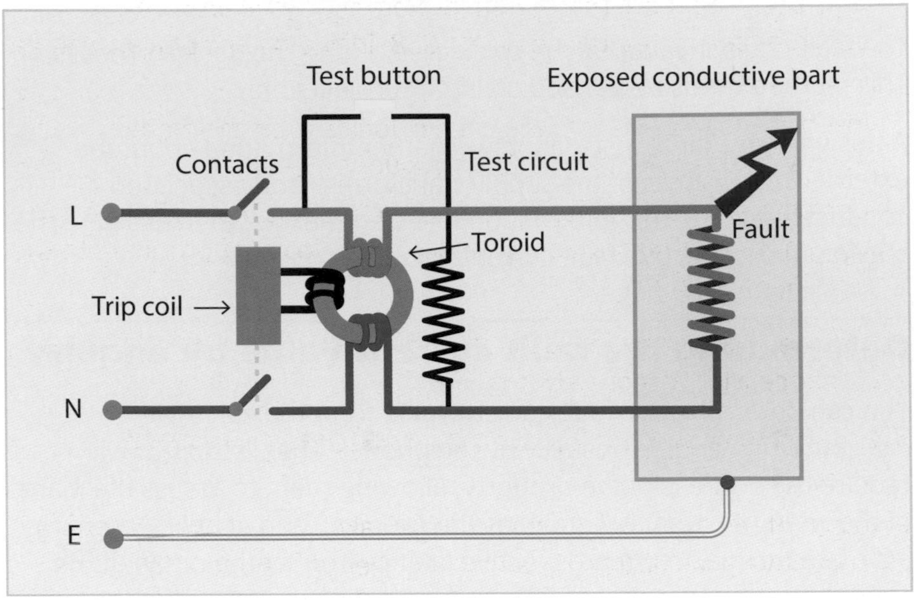

A line to earth (cpc) fault

In the presence of this fault, the RCD would detect an imbalance and the linked switch would open, as shown in the diagram.

The operating times of RCDs must be tested where they are essential for disconnection (for compliance with Chapter 41 of BS 7671) or where they are installed as additional protection as specified in Chapter 41.

The test procedure

The table shows the maximum operating times for non-time delayed RCDs at 100% of their rated tripping current.

	(ms)
BS 4293	200
BS 61008	300
BS 61009 (RCBO)	300
BS 7288 (integral socket-outlet)	200

Disconnection times for RCDs (extract from Guidance Note 3: Inspection and Testing, IET)

ACTIVITY

Look again at the location plan of the small industrial unit in Appendix 1 and use the circuit information to evaluate the correct tests for testing for the ring final circuit Br-3. The RCD is intended to be used for additional protection in accordance with Regulation 411.3.3.

ASSESSMENT GUIDANCE

Remember, the RCD tester gives results in milliseconds. Remember to test on both the positive and negative half cycle and take the highest reading.

Setting up the test meter for testing an RCD

Method

1 Select the correct meter – the correct meter is the RCD tester.

2 Select the correct setting for the RCD under test – this would be, for example, 30 mA, 100 mA or 300 mA.

3 Select the correct scale – the correct scale would be, where applicable, ½ rated, 1× rated or 5× rated.

4 Check the leads for damage – ensure that the probe tips are protruding no more than 4 mm (preferably 2 mm) and are in accordance with Guidance Note GS38, including any plug-in lead for socket outlets.

5 Insert the leads – choose the correct location on the meter (following the manufacturer's instructions).

6 Correctly connect the leads – you can choose either the correct plug-in lead for testing socket outlets or fly leads appropriate to the terminals to be tested.

Remember that all test leads must be manufactured and maintained in accordance with Guidance Note GS38.

Carrying out the tests for residual current devices (RCDs)

For each of the tests, readings should be taken on both positive (+ve) and negative (–ve) half cycles and the longer of the operating times recorded on the Schedule of Test Results. Some instruments may show the half cycles as 0° and 180° or as two sine waves in opposite directions.

The tests that need to be carried out depend on the reason why the RCD is fitted:

- 5× rated must be used if the RCD is present for additional protection and to prove disconnection times stated in Chapter 41 of BS 7671.

- ½ rated tests are carried out to ensure that the RCD is not *too* sensitive.

- 1× rated tests are carried out to ensure that the RCD is working to the manufactured standard.

RCDs rated greater than 30 mA $I_{\Delta n}$, do not require the 5× test. The 5× test is carried out as prescribed in Regulation 415.1 of BS 7671 for additional protection against electric shock.

Where periodic testing of an RCD that provides additional protection is being carried out, do the 5× test first as this is the most important. Any test following this is likely to be fast as the mechanical parts have been exercised.

The table below illustrates the maximum durations allowed (in red) and some typical values of test results (in blue).

BS or BS EN device	RCD rated tripping current $I_{\Delta n}$	½ rated $I_{\Delta n}$	1× rated $I_{\Delta n}$	5× rated $I_{\Delta n}$
BS EN 61009	30 mA	No trip	300 ms (max)	40 ms (max)
Typical test results		+ve – no trip –ve – no trip	+ve 128 ms –ve 114 ms	+ve 23 ms –ve 14 ms

RCD integral test button

The RCD integral test button is used to verify the function (and proper lubrication) of the mechanical parts, ensuring that the switching mechanism within the RCD is working correctly. Usually, this button can be seen on the front of the RCD.

Manufacturers normally recommend that the test button is pressed every three months (quarterly). This test must be done *before* all of the other tests for initial verification. However, during periodic inspection and testing, the integral test button is pressed *after* the tests.

Any device that is left in one position for a long time may stick. As RCDs rely on small amounts of energy (mA) to bring about disconnection, a sticky RCD may not trip in the time required. Regular pressing of the test button will improve the response speed of an RCD. Testing the RCD *before* pressing the test button during a periodic inspection will give a good indication of whether the RCD is being tested frequently enough!

Time delay RCDs

Time delay S-type RCD operating durations can be found in the table below.

There is no 5× test for S-type RCDs.

▼ **Table 2.9** Operational tripping times for various RCDs

Device type	Non-time delayed maximum operating time at 100% rated tripping current, $I_{\Delta n}$ (ms)	With time delay operating time at 100% rated tripping current, $I_{\Delta n}$ (ms)	Notes
BS 4293	200	{(0.5 to 1.0) × time delay} + 200	
BS 61008	300	130 to 500	S type
BS 61009 (RCBO)	300	130 to 500	S type
BS 7288 (integral socket-outlet)	200	non-applicable	

Disconnection times for non-time delayed and time delayed RCDs (from Guidance Note 3: Inspection and Testing, IET)

PHASE SEQUENCE TESTING

Phase sequence testing was introduced in BS 7671: 2008 (17th Edition) as Regulation 612.12. It is a requirement to verify that phase sequence is maintained for multi-phase circuits within an installation. As with many other regulations, damage or danger could result if the regulation is not followed.

Assessment criteria

4.17 State the methods used to check for the correct phase sequence

4.18 Explain why having the correct phase sequence is important

Why the correct phase sequence test is important

With three-phase rotating machines, you need to ensure that the direction of supply is correct. Starting and running a motor in the wrong direction could be disastrous. For example, consider what might happen if a unidirectional production process, pump motor or extraction system were to work in the wrong direction; production would go backwards, the pump would flood and the extractor would blow air in. Any of these could be very dangerous.

For the origin of the supply and all distribution boards, the phase rotation sequence must be correct. If these are all common, the fixed wiring within the building can be installed to maintain this phase sequence.

To keep the three-phase rotation common throughout, you must ensure that isolators, machine supplies and three-phase socket outlets are all following the same phase sequence.

ACTIVITY

Where would be the best place to test the phase sequence for a three-phase machine?

Equipment that requires phase sequence testing

The parts of an electrical installation that require phase sequence testing are:

- the origin of the supply
- main switches and associated distribution boards
- all sub-mains distribution boards
- three-phase and earth socket outlets (four-pin)
- three-phase, neutral and earth socket outlets (five-pin)
- three-phase isolators
- three-phase motor starters
- three-phase machine supplies and switchgear.

Do not attempt to test any equipment that may cause danger or risk of injury, such as motor terminals, or vibrating or moving machinery.

Methods used to check the correct phase sequence

Phase sequence testing can be checked by using a rotating disc or indicator lamp type of test meter that physically shows the rotation of the system or using the electronic LCD-type display shown below.

Rotating LCD display examples

Alternatively, the meter used could simply indicate the sequence of the three phases as shown below.

Phase rotation readout examples

Both types of phase sequence indicator can also be used to verify phase sequence or direction of rotation.

The test meter used should be supplied with test leads and probes in accordance with Guidance Note GS38.

The test meter will also require a plug-in facility for socket outlets. This plug-in facility must have extension adapters for different socket outlet ratings, ie 16 A, 32 A, 63 A, etc. Also, four-pin and five-pin socket-outlet adaptor variations may be required to test the installation socket outlets.

THE NEED FOR FUNCTIONAL TESTING

Functional testing must be carried out on equipment to ensure that it is working correctly. All intended functional switching devices must be capable of making and breaking their intended loads in accordance with the design and/or the functional switching procedures. An example of a functional test is RCD functional testing, described on pages 103–106 and carried out by using the integral test button.

Items that require functional testing

Equipment such as switchgear and controlgear assemblies, drives, controls and interlocks must be subjected to a functional test to show that they are properly mounted, adjusted and installed in accordance with the relevant requirements of BS 7671.

Switchgear and isolators

Functional testing is required for all equipment that may be relied on for protection against electric shock. For example, it is essential that a firefighters' switch is functioning correctly in an emergency situation.

Isolators must be located correctly in the electrical system and be fully functional. Many isolators are not designed to make and break under load conditions. Where this is the case, such isolators should not be located in positions where they are likely to be used as switches under load conditions.

Control switches and interlocks

Start and stop switches, interlocks and guard switches are examples of functional switching to ensure that production processes can be stopped immediately should danger arise.

Control circuits should be designed, arranged and protected in such a way as to prevent equipment from operating inadvertently in the event of a fault. It is good practice for control systems to be as simple as possible, while being fail-safe.

Although semiconductors can't be used as a means of isolation, they may be used for functional switching purposes.

RCDs and RCBOs

These devices must be functionally tested by carrying out the tests recommended by BS 7671 and associated guidance notes, including by pressing the integral test button marked 'T'.

Dimmer switches and speed controls

Dimmer switches and speed controls are not just simple on–off devices. These functional switches allow variable control of the luminance or the speed of a process respectively.

Motor control

Some motors use reverse-current braking and, where such reversal might result in danger, measures should be taken to prevent continued reversal after the driven parts come to a standstill at the end of the braking period. Further, where safety is dependent on the motor operating in the correct direction, means should be provided to *prevent* reverse operation. Where motor control systems incorporate overload devices, these devices should be checked to ensure the settings are correct and re-sets are correctly set to manual or automatic.

Circuit breakers

All circuit breakers should be checked for correct operation to ensure that they are opening and closing their intended circuits. Circuit protective devices are not intended for frequent load switching.

Firefighters' switch

Typical isolating switch

Emergency stop buttons

Residual current device

KEY POINT

All isolation and control switching must be checked for correct operation.

ACTIVITY

Make a list of all functional tests which need to be carried out, based on the location plan of a small industrial unit in Appendix 1.

Assessment criteria

4.20 Specify the methods used for verification of voltage drop

4.21 State the cause of voltage drop in an electrical installation

SmartScreen Unit 307

Handout 39

VOLTAGE DROP THEORY AND CALCULATION

Voltage drop is the product of the circuit conductor resistances and the maximum load current that the circuit conductors are expected to carry. The manufacturer of equipment normally states the minimum voltage necessary to ensure that the equipment functions correctly. Where no such limits are stated, values given in BS 7671 are considered suitable. The percentage voltage drop values given in BS 7671 are shown in the table.

TABLE 4Ab – Voltage drop

		Lighting	Other uses
(i)	Low voltage installations supplied directly from a public low voltage distribution system	3%	5%

Maximum voltage drop parameters (from BS 7671:2008 (2011))

Voltage drop calculation

The following information is required to establish the voltage drop by calculation.

- cable length (L)
- design current (I_b)
- value of millivolts per ampere per metre (mV/A/m) as given in Appendix 4 of BS 7671, which equates to the conductor resistance at full operating temperatures.

Using the values above, voltage drop could be determined by:

$$\frac{\text{mV/A/m} \times L \times I_b}{1000} = V$$

To use an example, assume a power circuit has the following values:

- cable length = 22 m
- design current = 13 A
- mV/A/m = 18 (taken from Table 4D1B below using column 3 for a 2.5mm² cable, reference method B)

The table on page 111 (top) is an extract from Table 4D1B, BS 7671. It shows the millivolt drop per ampere per metre (mV/A/m) for given cable types and insulations. When using the tables in BS 7671, the correct table should be used, depending on the cable type and material used for insulation of the conductors.

ACTIVITY

Cable calculations can be very time-consuming for large installations. Research the internet to find suitable IT software for the task.

VOLTAGE DROP (per ampere per metre):

Conductor cross-sectional area	2 cables, d.c.	2 cables, single-phase a.c.		
		Reference Methods A & B (enclosed in conduit or trunking)	Reference Methods C & F (clipped direct, on tray or in free air)	
			Cables touching	Cables spaced*
1	2	3	4	5
(mm²)	(mV/A/m)	(mV/A/m)	(mV/A/m)	(mV/A/m)
1	44	44	44	44
1.5	29	29	29	29
2.5	18	18	18	18
4	11	11	11	11
6	7.3	7.3	7.3	7.3
10	4.4	4.4	4.4	4.4
16	2.8	2.8	2.8	2.8

Voltage drop values extracted from Table 4D1B of BS 7671:2008 (2011)

So the actual voltage drop for the example would be:

$$\frac{mV/A/m \times L \times I_b}{1000} = V$$

$$\frac{18 \times 22 \times 13}{1000} = 5.15 \text{ V}$$

Maximum volt drop allowed for a power circuit = 5% of 230 V = 11.5 V (taken from Table 4Ab below).

As the value of 5.15 V is well below 11.5 V, this is satisfactory.

TABLE 4Ab – Voltage drop

	Lighting	Other uses
(i) Low voltage installations supplied directly from a public low voltage distribution system	3%	5%

Maximum voltage drop parameters (from BS 7671:2008 (2011))

Voltage drop verification and testing

Voltage drop problems are quite rare but the inspector should be aware that long runs and/or high currents can sometimes cause voltage drop problems.

There are two recognised methods of testing voltage drop:

■ measuring with voltage and current test equipment

■ measuring resistance.

ACTIVITY

What would be the maximum allowable voltage drop for a 400 V three-phase power circuit?

KEY POINT

Verification of voltage drop is not normally required during initial verification. This should be calculated during the design process, but initial verification does suggest that it should be considered.

ASSESSMENT GUIDANCE

Low voltage problems are rare these days. Low voltage might, however, be noticed in remote rural areas at the end of a long line, when a heavy load is applied. Remember that 230 V is only the nominal voltage.

Method 1: Measuring with voltage and current test equipment

This method is *not* recommended due to the need to work on or near live parts and having exposed live terminals.

Measurement of voltage drop within an installation is not practical as this would mean simultaneously measuring the instantaneous voltage at both the origin and at the end point of the circuit. This would have to be done with the circuit operating under fully loaded conditions.

Method 2: Measuring resistance

Practical testing is more complicated and less accurate than calculating voltage drop by analysing the design criteria and using the calculation format explained on page 111.

If you do decide to test the resistance of the cable, *isolation procedures* must be followed and a risk assessment should be carried out.

How to test a (single-phase) radial circuit

To test a (single-phase) radial circuit you will need to measure the resistance of the line and neutral conductors (in ohms). This can be done using the same procedure as is used to measure R_1+R_2 but, instead of measuring the cpc resistance (R_2), you measure the neutral resistance (R_n).

Once a value of R_1+R_n has been obtained by measurement, it must be adjusted for temperature. If the cable has a maximum operating temperature of 70 °C, the resistance value should be adjusted from that measured at the ambient temperature, to that at 70 °C.

As resistance changes by 2% per 5 °C change in temperature, you can determine the resistance increase using the following table.

Ambient temperature at time of test (°C)	Temperature increase to maximum operating temperature (°C)	% increase	Factor to be applied
10	60	24	1.24
15	55	22	1.22
20	50	20	1.20
25	45	18	1.18
30	40	16	1.16
35	35	14	1.14
40	30	12	1.12

Multipliers for adjusting resistance of conductors such as thermoplastic 70 °C cable

For example, the $R_1 + R_n$ for a radial power circuit was 0.22 Ω at an ambient temperature of 15 °C. If the cable has an operating temperature of 70 °C, we can determine the resistance under operating conditions:

$$0.22 \times 1.22 = 0.27 \text{ at } 70 °C$$

As voltage drop is determined using Ohm's law, we need to consider the full load current. For this example, we will assume full load current is 24 A.

$$\text{So as } I \times R = V$$
$$24 \text{ A} \times 0.27 \text{ Ω} = 6.48 \text{ V}$$

TABLE 4Ab – Voltage drop

		Lighting	Other uses
(i)	Low voltage installations supplied directly from a public low voltage distribution system	3%	5%

Maximum voltage drop parameters (from BS 7671:2008 (2011))

The calculated voltage drop is 6.48 V. As the maximum permitted voltage drop for a radial power circuit is 5% of 230 V (= 11.5 V), this value is acceptable.

APPROPRIATE PROCEDURES FOR DEALING WITH CUSTOMERS AND CLIENTS

As soon as initial verification and certification is complete, the client will want to start using the electrical equipment. It is essential that the information given to the client is correctly explained. The relevant information and requirements are outlined below and in the following pages.

Commissioning and certification

Client awareness during handover

All installations differ in complexity, from domestic to large industrial. In every case, as part of the handover, the client or customer must be made aware of the safe and correct use of the electrical installation and all of its controls. This can be achieved by giving a tour of the installation to explain any complexities.

Any user manuals and information booklets must be given to the client. In the case of complex installations or commercial and industrial installation, this would normally form the basis of an operations and maintenance manual (O&M Manual).

Certification presented during handover

After initial verification and commissioning is complete, the certification paperwork must be handed over to the client or the person ordering the work.

These documents are:

- Electrical Installation Certificate
- Schedule of Inspections
- Generic Schedule of Test Results.

All three documents must be complete and satisfactory before the client uses the electrical system.

Where minor additions or alterations to a single existing circuit have taken place, this would be recorded on a Minor Electrical Installation Works Certificate.

In every case, the certification cannot be completed and handed over until all work is satisfactory and complies with BS 7671.

Identification of circuits and switches

Distribution board and consumer unit protective devices must be identified correctly and information must be available for the client in accordance with Regulation 514.9.1 of BS 7671.

A durable copy of the circuit details relating to the distribution board (DB) or consumer control unit (CCU) should be placed within or adjacent to each DB or CCU. Diagrams, plans and labelling of isolators and switchgear must be accurate.

Why the client requires certification

There are many reasons for certification. It is required:

- in accordance with British Standards (BS 7671)
- to meet health and safety regulations in the workplace (Electricity at Work Regulations 1989)
- for domestic premises as part of Building Regulations
- for building insurance
- for proof of compliance in the event of any employee and personal insurance claims
- before letting a property (Landlord and Tenant Act 1985)
- for local authority licensing requirements, such as in the case of obtaining a public entertainments license or fire certification.

Why certification must be maintained

After the electrical installation certification is handed over, the documents must be regularly updated to record alterations and associated additional certification, and as part of the maintenance of the system.

Future periodic testing can be carried out and historical test results can be compared to ensure that there are no major changes or patterns of deterioration within the electrical system.

If the client is selling the property, the purchaser will require the certification, as well as any subsequent reports.

ACTIVITY

It is a worthwhile exercise to for you to put yourself in the place of a client. Consider this scenario: you are the owner of a new hotel and handover of your newly installed electrical installation is due to take place. List the documents relating to the electrical installation that should be given to you during the handover process.

Understand the procedures and requirements for the completion of electrical installation certificates and related documentation

DOCUMENTATION AND CERTIFICATION FOR INITIAL VERIFICATION

This section provides guidance on completing the necessary forms and certificates associated with initial verification.

New installations

Following an initial verification, or an addition or alteration to an existing installation, an Electrical Installation Certificate must be completed and issued, together with Inspection Schedules and Test Result Schedules.

There are two options for the Electrical Installation Certificate, as shown on Form 1 (page 116) and Form 2 (page 117).

Form 1: Short Form of Electrical Installation Certificate

This form is used when *one* person is responsible for the design, construction, inspection and testing of an installation. This is normally used for small installations, such as domestic dwellings, where one person carries out all of the roles. Even if the person has only selected cable cross-sectional areas using tables in the IET On-Site Guide, they must take responsibility for design.

Form 2: Electrical Installation Certificate

This form is used when the design, construction and the inspection are carried out by different groups of people (it is a three-signatory version of the form). This form is normally used for larger projects where the installation has been designed by a consultant or a member of the architects' team, has been installed by many people working for a contractor and has been inspected by a specialist inspector – who is either part of the installation organisation, or is nominated by the client.

Forms 3 and 4: Schedule of Inspections and Generic Schedule of Test Results

Whichever of the two Electrical Installation Certificates is used, other forms are required to accompany the Certificate: Form 3 (Schedule of Inspections) Form 4 (Schedule of Test Results).

Completed samples of Forms 3 and 4 from Guidance Note 3 are shown on pages 118 and 119.

Assessment criteria

5.1 Explain the purpose of and relationship between:

- an electrical installation certificate
- a minor electrical installation works certificate
- schedule of inspections
- schedule of test results

SmartScreen Unit 307
Handout 42

ASSESSMENT GUIDANCE

Remember to complete all parts of the certificates and forms as appropriate. If any parts are left incomplete, this may mean the forms are not valid.

Form 1

Form No: 505513......./1

ELECTRICAL INSTALLATION CERTIFICATE
(REQUIREMENTS FOR ELECTRICAL INSTALLATIONS - BS 7671 (IET WIRING REGULATIONS))

DETAILS OF THE CLIENT Mr T Brown
32 South St
Anytown, Surrey
Post Code: TO1 1ZZ

INSTALLATION ADDRESS The Coffee Bean
31 Station Road
Anytown, Surrey
Post Code: TO3 2YF

DESCRIPTION AND EXTENT OF THE INSTALLATION Tick boxes as appropriate

Description of installation:
Re-wire of ground floor, on change of use.

New installation	☑
Addition to an existing installation	☐
Alteration to an existing installation	☐

Extent of installation covered by this Certificate:
Complete electrical re-wire of refurbished premises, on change of use from offices to café/snack bar.

(Use continuation sheet if necessary) see continuation sheet No: ......

FOR DESIGN, CONSTRUCTION, INSPECTION & TESTING
I being the person responsible for the design, construction, inspection & testing of the electrical installation (as indicated by my signature below), particulars of which are described above, having exercised reasonable skill and care when carrying out the design, construction, inspection & testing hereby CERTIFY that the said work for which I have been responsible is to the best of my knowledge and belief in accordance with BS 7671:2008, amended to 2011...... (date) except for the departures, if any, detailed as follows:

Details of departures from BS 7671 (Regulations 120.3 and 133.5):
None

The extent of liability of the signatory is limited to the work described above as the subject of this Certificate.
Signature: *W Hastings* Date: 21-Jan-2011 Name (IN BLOCK LETTERS): W HASTINGS

Company Hastings Electrical
Address: 21 The Arches
Anytown, Surrey Postcode: TO2 9YY Tel No: 01022 999999

NEXT INSPECTION
I recommend that this installation is further inspected and tested after an interval of not more than ......5...... years/months.

SUPPLY CHARACTERISTICS AND EARTHING ARRANGEMENTS Tick boxes and enter details, as appropriate

Earthing arrangements	Number and Type of Live Conductors		Nature of Supply Parameters	Supply Protective Device Characteristics
TN-C ☐	a.c. ☑	d.c. ☐	Nominal voltage, U/U₀ $^{(1)}$ .....230. V	Type BS 1361 Fuse
TN-S ☐	1-phase, 2-wire ☑	2-wire ☐	Nominal frequency, f $^{(1)}$ ....50. Hz	
TN-C-S ☑	1-phase, 3-wire ☐	3-wire ☐	Prospective fault current, I$_{pf}$ $^{(2)}$ ..9.0. kA	Rated current ......100..A
TT ☐	2-phase, 3-wire ☐	other ☐	External loop impedance, Z$_e$ $^{(2)}$ 0.28 ⊠	
IT ☐	3-phase, 3-wire ☐			
Other sources of supply (to be detailed on attached schedules) ☐	3-phase, 4-wire ☐			
	Confirmation of supply polarity ☑		*(Note: (1) by enquiry, (2) by enquiry or by measurement)*	

Form 1

Form No: 505513......./1

PARTICULARS OF INSTALLATION REFERRED TO IN THE CERTIFICATE Tick boxes and enter details, as appropriate

Means of Earthing

Distributor's facility	☑
Installation earth electrode	☐

Maximum Demand

Maximum demand (load) .....80. ~~kVA~~ / Amps Delete as appropriate

Details of Installation Earth Electrode *(where applicable)*

Type (e.g. rod(s), tape etc) N/A	Location N/A	Electrode resistance to Earth N/A ⊠

Main Protective Conductors

Earthing conductor:	material Copper	csa 16 mm²	Continuity and connection verified ☑
Main protective bonding conductors	material Copper	csa 10 mm²	Continuity and connection verified ☑
To incoming water and/or gas service ☑	To other elements: N/A		

Main Switch or Circuit-breaker

BS, Type and No. of poles BS EN 60947-3 (2 pole) Current rating .....100.A Voltage rating .....230.. V

Location Services cupboard adjacent rear exit Fuse rating or setting.......N/A.A

Rated residual operating current I$_{\Delta n}$ = ...N/A mA, and operating time of ......N/A ms (at I$_{\Delta n}$) (applicable only where an RCD is suitable and is used as a main circuit-breaker)

COMMENTS ON EXISTING INSTALLATION (in the case of an addition or alteration see Section 633):
Not Applicable.

SCHEDULES
The attached Schedules are part of this document and this Certificate is valid only when they are attached to it.
.....1..... Schedules of Inspections and .....1..... Schedules of Test Results are attached.
(Enter quantities of schedules attached).

Form 1: Short form of Electrical Installation Certificate (always check you are using the latest forms, as found on the IET website: http://electrical.theiet.org)

Form 2   Form No: .....SSSS13.........../2

PARTICULARS OF SIGNATORIES TO THE ELECTRICAL INSTALLATION CERTIFICATE

Designer (No 1)
Name: ......D.Jones........ Company: .....T.&. Electrical Design Partnership....
Address: ....23.High Street........ Postcode: ....SL10.0YY..... Tel No: ....01000.999999....
....Somestown, Berks.......

Designer (No 2) (if applicable)
Name: .... Company: ....
Address: .... Postcode: .... Tel No: ....

Constructor
Name: ......T.Smith........ Company: ....T.Smith Electrical Installations....
Address: ...Unit 3a, Somestown Ind. Estate.... Postcode: ....SL3.0XX.... Tel No: ....01000.888888....
....Somestown, Berks.......

Inspector
Name: .....G.Wilson........ Company: ....Wilson and Sons....
Address: ...11 Oaktree Row.... Postcode: ....SL2.0WW.... Tel No: ....01000.777777....
....Somestown, Berks.......

SUPPLY CHARACTERISTICS AND EARTHING ARRANGEMENTS *Tick boxes and enter details, as appropriate*

Earthing arrangements	Number and Type of Live Conductors	Nature of Supply Parameters	Supply Protective Device Characteristics
TN-C ☐	a.c. ☑ d.c. ☐	Nominal voltage, U/U₀ (1) ....230.... V	Type: ....BS.88.3....
TN-S ☐	1-phase, 2-wire ☐ 2-wire ☑	Nominal frequency, f (1) ....50.... Hz	
TN-C-S ☑	1-phase, 3-wire ☐ 3-wire ☐	Prospective fault current, Ipf (2) ....1.41.... kA	
TT ☐	2-phase, 3-wire ☐ other ☐	External loop impedance, Ze (2) ....0.34Ω....	Rated current....80.... A
IT ☐	3-phase, 3-wire ☐	*(Note: (1) by enquiry, (2) by enquiry or by measurement)*	
Other sources ☐ (to be detailed on attached schedules)	3-phase, 4-wire ☐		
	Confirmation of supply polarity ☑		

PARTICULARS OF INSTALLATION REFERRED TO IN THE CERTIFICATE *Tick boxes and enter details, as appropriate*

Means of Earthing	Maximum Demand
Distributor's facility ☑	Maximum demand (load) ....68.... kVA / Amps *Delete as appropriate*

Details of Installation Earth Electrode (*where applicable*)
Installation earth electrode ☐
Type (e.g. rod(s), tape etc) ....N/A....   Location ....N/A....   Electrode resistance to Earth ....N/A.... Ω

Main Protective Conductors
Earthing conductor: material ....Copper.... csa ....16.... mm²   Continuity and connection verified ☑
Main protective bonding conductors material ....Copper.... csa ....16.... mm²   Continuity and connection verified ☑
To incoming water and/or gas service ☑   To other elements: ....N/A....

Main Switch or Circuit-breaker
BS, Type and No. of poles ....BS EN 60947.3.(2 pole).... Current rating ....100....A   Voltage rating ....400.... V
Location ....Office Switch Consumer Unit....   Fuse rating or setting ....N/A.... A
Rated residual operating current Iₙ = ...N/A.... mA, and operating time of ....N/A.... ms (at Iₙ) *(applicable only where an RCD is suitable and is used as a main circuit breaker)*

COMMENTS ON EXISTING INSTALLATION (in the case of an addition or alteration see Section 633):
....N/A....

SCHEDULES
The attached Schedules are part of this document and this Certificate is valid only when they are attached to it.
....1.... Schedules of Inspections and ....1.... Schedules of Test Results are attached.
(Enter quantities of schedules attached)

Form 2   Form No: .....SSSS13.........../2

ELECTRICAL INSTALLATION CERTIFICATE
(REQUIREMENTS FOR ELECTRICAL INSTALLATIONS - BS 7671 [IET WIRING REGULATIONS])

DETAILS OF THE CLIENT ....Mr.D Roberts....
....23 Acacia Avenue....
....Somestown, Berks....   Post Code: ....SL0.0LT....

INSTALLATION ADDRESS ....Unit 3, The Quadrant....
....Somestown Business Park....
....Somestown, Berks....   Post Code: ....SL1.0ZZ....

DESCRIPTION AND EXTENT OF THE INSTALLATION Tick boxes as appropriate

Description of installation: Commercial Office

Extent of installation covered by this Certificate: Full new installation

(Use continuation sheet if necessary)   see continuation sheet No. ....

New installation ☑
Addition to an existing installation ☐
Alteration to an existing installation ☐

FOR DESIGN
I/We being the person(s) responsible for the design of the electrical installation (as indicated by my/our signatures below), particulars of which are described above, having exercised reasonable skill and care when carrying out the design hereby CERTIFY that the design work for which I/we have been responsible is to the best of my/our knowledge and belief in accordance with BS 7671:2008, amended to ....2011.... (date) except for the departures, if any, detailed as follows:
Details of departures from BS 7671 (Regulations 120.3 and 133.5):
None   N/A

The extent of liability of the signatory or the signatories is limited to the work described above as the subject of this Certificate.

For the DESIGN of the installation: **(Where there is mutual responsibility for the design)
Signature: ....D.Jones.... Date: ...15/08/2013.... Name (IN BLOCK LETTERS): ....D.JONES.... Designer No 1
Signature: ....N/A.... Date: .... Name (IN BLOCK LETTERS): ....N/A.... Designer No 2**

FOR CONSTRUCTION
I/We being the person(s) responsible for the construction of the electrical installation (as indicated by my/our signatures below), particulars of which are described above, having exercised reasonable skill and care when carrying out the construction hereby CERTIFY that the construction work for which I/we have been responsible is to the best of my/our knowledge and belief in accordance with BS 7671:2008, amended to ....2011.... (date) except for the departures, if any, detailed as follows:
Details of departures from BS 7671 (Regulations 120.3 and 133.5):
None   N/A

The extent of liability of the signatory is limited to the work described above as the subject of this Certificate.

For CONSTRUCTION of the installation:
Signature: ....T.Smith.... Date: ....15/08/2013.... Name (IN BLOCK LETTERS): ....T.SMITH....

FOR INSPECTION & TESTING
I/We being the person(s) responsible for the inspection & testing of the electrical installation (as indicated by my/our signatures below), particulars of which are described above, having exercised reasonable skill and care when carrying out the inspection & testing hereby CERTIFY that the work for which I/we have been responsible is to the best of my/our knowledge and belief in accordance with BS 7671:2008, amended to ....2011.... (date) except for the departures, if any, detailed as follows:
Details of departures from BS 7671 (Regulations 120.3 and 133.5):
None   N/A

The extent of liability of the signatory is limited to the work described above as the subject of this Certificate.

For INSPECTION AND TESTING of the installation:
Signature: ....G.Wilson.... Date: ....15/08/2013.... Name (IN BLOCK LETTERS): ....G.WILSON....

NEXT INSPECTION
I/We the designer(s), recommend that this installation is further inspected and tested after an interval of not more than ....5.... years/months.

Form 2: Electrical Installation Certificate (three-signatory version) (always check you are using the latest forms, as found on the IET website: http://electrical.theiet.org)

Form 3   Form No: ...*SSSS13*..../3

SCHEDULE OF INSPECTIONS (for new installation work only)

Methods of protection against electric shock	Prevention of mutual detrimental influence

Methods of protection against electric shock

Both basic and fault protection:

- [✓] (i)   SELV (note 1)
- [N/A] (ii)   PELV
- [N/A] (iii)   Double insulation
- [N/A] (iv)   Reinforced insulation

Basic protection: (note 2)

- [✓] (i)   Insulation of live parts
- [✓] (ii)   Barriers or enclosures
- [N/A] (iii)   Obstacles (note 3)
- [N/A] (iv)   Placing out of reach (note 4)

Fault protection:

(i)   Automatic disconnection of supply:

- [✓]   Presence of earthing conductor
- [✓]   Presence of circuit protective conductors
- [✓]   Presence of protective bonding conductors
- [✓]   Presence of supplementary bonding conductors
- [N/A]   Presence of earthing arrangements for combined protective and functional purposes
- [N/A]   Presence of adequate arrangements for other sources, where applicable
- [N/A]   FELV
- [✓]   Choice and setting of protective and monitoring devices (for fault and/or overcurrent protection)

(ii)   Non-conducting location: (note 5)

- [N/A]   Absence of protective conductors

(iii)   Earth-free local equipotential bonding: (note 6)

- [N/A]   Presence of earth-free local equipotential bonding

(iv)   Electrical separation: (note 7)

- [N/A]   Provided for **one item** of current-using equipment
- [N/A]   Provided for **more than one item** of current-using equipment

Additional protection:

- [✓]   Presence of residual current devices(s)
- [✓]   Presence of supplementary bonding conductors

Prevention of mutual detrimental influence

- [✓]   (a)   Proximity to non-electrical services and other influences
- [✓]   (b)   Segregation of Band I and Band II circuits or use of Band II insulation
- [N/A]   (c)   Segregation of safety circuits

Identification

- [✓]   (a)   Presence of diagrams, instructions, circuit charts and similar information
- [✓]   (b)   Presence of danger notices and other warning notices
- [✓]   (c)   Labelling of protective devices, switches and terminals
- [✓]   (d)   Identification of conductors

Cables and conductors

- [✓]   Selection of conductors for current-carrying capacity and voltage drop
- [✓]   Erection methods
- [✓]   Routing of cables in prescribed zones
- [N/A]   Cables incorporating earthed armour or sheath, or run within an earthed wiring system, or otherwise adequately protected against nails, screws and the like
- [N/A]   Additional protection provided by 30 mA RCD for cables concealed in walls (where required in premises not under the supervision of a skilled or instructed person)
- [✓]   Connection of conductors
- [✓]   Presence of fire barriers, suitable seals and protection against thermal effects

General

- [✓]   Presence and correct location of appropriate devices for isolation and switching
- [✓]   Adequacy of access to switchgear and other equipment
- [N/A]   Particular protective measures for special installations and locations
- [✓]   Connection of single-pole devices for protection or switching in line conductors only
- [✓]   Correct connection of accessories and equipment
- [N/A]   Presence of undervoltage protective devices
- [✓]   Selection of equipment and protective measures appropriate to external influences
- [✓]   Selection of appropriate functional switching devices

Inspected by ....*G.Wilson*...   Date ....*15/08/2013*...

NOTES:

✓   to indicate an inspection has been carried out and the result is satisfactory
N/A   to indicate that the inspection is not applicable to a particular item
An entry must be made in every box.

1.   SELV An extra-low voltage system which is electrically separated from Earth and from other systems. The particular requirements of the Regulations must be checked (see Section 414)
2.   Method of basic protection - will include measurement of distances where appropriate
3.   Obstacles - only adopted in special circumstances (see Regulations 416.2 and 417.2)
4.   Placing out of reach - only adopted in special circumstances (see Regulation 417.3)

5.   Non-conducting locations - not applicable in domestic premises and requiring special precautions (see Regulation 418.1)
6.   Earth-free local equipotential bonding - not applicable in domestic premises, only used in special circumstances (see Regulation 418.2)
7.   Electrical separation (see Section 413 and Regulation 418.3)

Page *3*... of .*4*...

Form 3: Schedule of Inspections (always check you are using the latest forms, as found on the IET website: http://electrical.theiet.org)

Form 4

Form No: ......1235........./4

GENERIC SCHEDULE OF TEST RESULTS

DB reference no ....Commercial office....
Location ..Unit 3, The Quadrant, SL1 0ZZ
Zs at DB (Ω) .....0.34....
Ipf at DB (kA) .....1.41....
Correct supply polarity confirmed ☑
Phase sequence confirmed (where appropriate) [N/A]

Details of circuits and/or installed equipment vulnerable to damage when testing ....Downlighters spots - electronic SELV transformers.

Details of test instruments used (state serial and/or asset numbers)
Continuity....Megin multi-function. 10563.
Insulation resistance ...."....
Earth fault loop impedance ...."....
RCD ...."....
Earth electrode resistance ......N/A.

Tested by:
Name (Capitals) ....G WILSON....
Signature ....G Wilson.... Date ....15/08/2013....

Page 4 ... of ...4

Circuit number	Circuit Description	Overcurrent device BS (EN)	type	rating (A)	breaking capacity (kA)	Conductor details Reference Method	Live (mm²)	cpc (mm²)	Ring final circuit continuity (Ω) r1 (line)	rn (neutral)	r2 (cpc)	Continuity (Ω)(R1+R2) or R2 (R1+R2)*	R2	Insulation Resistance (MΩ) Live-Live	Live-E	Polarity	Zs (Ω)	RCD (ms) @ ΔN	@ 5 ΔN	Test button / functionality	Remarks (continue on a separate sheet if necessary)
1	Socket outlets - hub	60898	B	20	6	B	2.5	1.5	N/A	N/A	N/A	0.38	N/A	250	250	✓	0.75	N/A	N/A	N/A	✓ Checked for compliance
2	Socket outlets - wall	61009	B	32	6	B	2x2.5	2x1.5	0.62	0.62	1.02	0.41	N/A	250	250	✓	0.71	85	16	✓	- " -
3	Down lighter spots	60898	B	6	6	B	1.5	1.0	N/A	N/A	N/A	0.56	N/A	250	250	✓	0.83	N/A	N/A	N/A	- " -
4	General lighting	60898	B	10	6	B	1.5	1.0	N/A	N/A	N/A	0.48	N/A	250	250	✓	0.81	N/A	N/A	N/A	- " -
5	Water heater	60898	B	16	6	B	2.5	1.5	N/A	N/A	N/A	0.11	N/A	250	250	✓	0.45	N/A	N/A	N/A	- " -

* Where there are no spurs connected to a ring final circuit this value is also the (R1 + R2) of the circuit.

Form 4: Generic Schedule of Test Results (always check you are using the latest forms, as found on the IET website: http://electrical.theiet.org)

New circuits

For installation of one or more new circuits, certification and schedules are required:

- Form 1 or 2: Electrical Installation Certificate
- Form 3: Schedule of Inspections
- Form 4: Generic Schedule of Test Results.

Additions and alterations to existing circuits

When completing the Electrical Installation Certificate for additions or alterations to an existing electrical installation, you may need to comment on the existing installation. If there are apparent problems or deficiencies, you must inform the person ordering the work about any possible risks if remedial work is not carried out.

If the additions or alterations do not include the provision of a new circuit, the Form 5 certificate needs to be issued.

Form 5: Minor Electrical Installation Works Certificate

A sample of Form 5, taken from Guidance Note 3, is shown opposite. The Minor Electrical Installation Works Certificate can only be used if the minor electrical work does not include the provision of a new circuit. When replacing equipment and accessories, or adding to an existing final circuit, this form can be used.

ASSESSMENT GUIDANCE

A new socket added to an existing ring final circuit would need a Minor Electrical Installation Works Certificate. The same socket wired back to the consumer unit or distribution board would need an Electrical Installation Certificate plus Schedule of Inspections and Generic Schedule of Test Results.

KEY POINT

Remember that, if a new circuit is installed, you must use the Electrical Installation Certificate format.

ACTIVITY

What form is most likely to be used after the replacement of a faulty water heater?

Form 5   Form No: ....1234.../5

MINOR ELECTRICAL INSTALLATION WORKS CERTIFICATE
(REQU REMENTS FOR ELECTRICAL INSTALLATIONS - BS 7671 [IET W RING REGULATIONS])
To be used only for minor electrical work which does not include the provision of a new circuit

---

PART 1 Description of minor works

1. Description of the minor works **2 new lighting points to home office/bedroom 3 of dwelling.**

2. Location/Address **41 Larkspur Drive, Newtown. E. Sussex**

   Post Code **EA1 2BB**

3. Date minor works completed **27 Jan 2012**

4. Details of departures, if any, from BS 7671:2008, amended to ...**2011**..... (date)
   None. Electricity supplier's terminal equipment in need of attention. Cut-out fuse carrier cracked. Customer advised to contact supplier.

PART 2 Installation details

1. System earthing arrangement   TN-C-S ☑   TN-S ☐   TT ☐

2. Method of fault protection **ADS**

3. Protective device for the modified circuit   Type **BS EN 61009 Type B**   Rating ....**6**. A

Comments on existing installation, including adequacy of earthing and bonding arrangements (see Regulation 132.16):

Existing circuit not provided with additional protection by RCD. Lighting circuit MCB converted to RCBO in order adequately to provide protection against damage to cables in walls; Reg. 522.6.101 refers.

PART 3 Essential Tests
Earth continuity satisfactory ☑

Insulation resistance:
          Line/neutral ....**+299**. MΩ

          Line/earth ....**+299**. MΩ

          Neutral/earth.............................**+299**. MΩ

Earth fault loop impedance ....**1,2**. Ω

Polarity satisfactory ☑

RCD operation (if applicable). Rated residual operating current $I_{\Delta n}$ ...**30**. mA and operating time of ....**28**.ms (at $I_{\Delta n}$)

PART 4 Declaration

I/We CERTIFY that the said works do not impair the safety of the existing installation, that the said works have been designed, constructed, inspected and tested in accordance with BS 7671:2008 (ET Wiring Regulations), amended to **2011**........ (date) and that the said works, to the best of my/our knowledge and belief, at the time of my/our inspection, complied with BS 7671 except as detailed in Part 1 above.

Name: **G Thompson**

For and on behalf of: **T and G Electrical**

Address: **25 Whiteleaf Close**
Newtown
E Sussex
.....................................Post code. **EA4 5XX**

Signature: ....**G. Thompson**............

Position: **Proprietor**

Date: ....**27-Jan-2012**..............

Page 1 of 1

Form 5: Minor Electrical Installation Works Certificate (always check you are using the latest forms, as found on the IET website: http://electrical.theiet.org)

Assessment criteria

5.2 State the information that must be contained within:

- an electrical installation certificate
- a minor electrical installation works certificate
- schedule of inspections
- schedule of test results

INFORMATION CONTAINED WITHIN DOCUMENTATION AND CERTIFICATION

This section deals with the information that must be contained within the documentation and certification for initial verification.

Information contained within an Electrical Installation Certificate (Forms 1 and 2)

An Electrical Installation Certificate contains details of the person(s) responsible for the design, construction, inspection and testing of the installation.

It also identifies the:

- details of the installation covered by the certificate
- nature of the works; e.g. new, addition, alteration
- extent of the work covered by the certificate
- supply characteristics
- nature of supply parameters
- earthing arrangements
- maximum demand
- protective conductors and their cross-sectional area
- main switch type and standard
- comments on the condition of any existing electrical systems which are not covered by the certificate but are part of the general electrical installation.

Information contained within a Minor Electrical Installation Works Certificate (Form 5)

The minor works which have been undertaken must be clearly described on this certificate. The purpose of the certificate is to verify that the new works comply with BS 7671 and that the existing parts of the circuit and affected parts of the installation are suitable for the additional works and remain compliant.

The relevant provisions of Part 6 'Inspection and testing' of BS 7671 must be applied in full to all minor works. For example, where a socket outlet is added to an existing circuit it is necessary to:

- establish that the earthing contact of the socket outlet is connected to the main earthing terminal
- measure the insulation resistance of the circuit that has been added to, and establish that it complies with Table 61 of BS 7671

ACTIVITY

There is a danger that a like-for-like change (shower unit for shower unit) does not take into account a possible increase in load current. This may well require a new circuit, with consequent design considerations. What certification would be needed in such a case?

ASSESSMENT GUIDANCE

Don't forget to date the forms. They are not useful unless they are dated.

- measure the earth fault loop impedance to establish that the maximum permitted disconnection time is not exceeded

- check that the polarity of the socket outlet is correct

- verify the effectiveness of the RCD, if the new work is protected by an RCD.

Information contained within a Schedule of Inspections (Form 3)

A Schedule of Inspections contains boxes which need to be filled in with a tick (for compliance) or with 'n/a' for items which are not applicable. The inspection methods and processes have been addressed on pages 27–42.

This form must accompany the Electrical Installation Certificate.

The table below shows each tick box within the Schedule of Inspections, together with the questions that the inspector must ask while inspecting the installation. The table also provides some additional information to help the inspector with the inspection. Where aspects of the inspection need more detailed information the relevant BS 7671 Regulation number or section is given.

Schedule of Inspections (Form 3) – information and guidance

Methods of protection against electric shock	
Both basic and fault protection	
SELV	Does the installation contain SELV equipment such as extra-low voltage lighting? Is the transformer compliant with Regulation 414.3? SELV circuits must be separated from the source and earth. The equipment will not have a connection to earth.
PELV	Does the installation contain any PELV equipment? This is where an item of equipment is supplied using extra-low voltage but may be earthed using the supply circuit cpc. An example may be a server system for data where the cables and equipment casings are earthed but the circuit is separated.
Double Insulation	Does the installation or part of an installation rely on double insulation as a method of shock protection, instead of ADS? It is tempting to tick this box where the supply tails are double insulated, but this case is not relevant to this box as the tails only have double insulation for mechanical protection. See the requirements of Section 413. It is rare for this box to be applicable.

Reinforced insulation	Does the installation or part of the installation rely on reinforced insulation as a method of shock protection, instead of ADS?
	See the requirements of Section 413.
	It is rare for this box to be applicable.

Basic protection

Insulation of live parts	Is basic shock protection provided by insulation of live parts (such as conductors within insulated cables)?
	This box is relevant to *all* electrical installations and is addressed through inspection.
	See Section 416.1.
Barriers and enclosures	Do all barriers and enclosures (such as switch or socket outlet boxes and DBs etc.) have suitable IP (international protection) ratings? Are they secured by a tool or key to stop people accidently touching live connections?
	See Section 416.2.
Obstacles	Are obstacles present and sufficient?
	Obstacles are rare in installations but aim to stop people touching live parts. They are not necessarily secured by a tool or key and do not have an IP rating like an enclosure.
	An example could be where open switch contacts are behind a safety rail within a restricted switch room.
	See Section 417.2.
Placing out of reach	Is 'placing out of reach' used?
	Placing out of reach, such as in a bare overhead conductor or bus bar system, is rare in installations.
	The requirements of Section 417.3 must be met.

Fault protection

(i) Automatic disconnection of supply

Presence of earthing conductor	Is the earthing conductor present, connecting the MET to the means of earthing? Is it continuous and correctly sized in accordance with Chapter 54? This is applicable in most installations.
Presence of cpc	As required by 411.3.1.1, is a cpc present in all circuits where automatic disconnection of supply (ADS) is the protective measure? Is the cpc suitably sized in accordance with Chapter 54? This is applicable in most installations.
Presence of main protective bonding conductors	Does the installation contain any extraneous parts as listed in Regulation 411.3.1.2? If so, they must be bonded to the main earthing terminal (MET) by a conductor in accordance with Chapter 54. This is applicable in most installations.
Presence of supplementary bonding conductors	Is this a special location, such as a bathroom, where supplementary bonding is required to link all exposed and extraneous parts? Supplementary bonding is becoming less common with the use of RCD protection. However, it is recommended that supplementary bonding is installed in some situations. For example, where RCDs are not maintained in accordance with the manufacturer's recommendations (such as by pressing the test button quarterly), there is no guarantee that the RCD will function correctly under fault conditions. Therefore, in a domestic situation where ordinary persons are present and where the installation is not under effective supervision, it may be prudent to install supplementary protective bonding conductors in locations containing a bath or a shower.
Presence of earthing arrangements for combined protective and functional purposes	Is there a protective conductor that also acts as a zero volt reference for equipment, such as telecoms, to function? This is not common in most installations. The cable should be coloured cream to identify this situation.

Presence of adequate arrangements for other sources where applicable	Is there adequate switching and protection given where additional sources of energy exist, such as PV sources or standby generators/uninterruptable power supplies (UPS)? Further detail exists in Chapter 55.
FELV	Is FELV used to reduce voltage to ELV for the purpose of function, such as in machine controls? This is not common in installations. The requirements of ADS must apply to the source circuit. See Section 411.7.
Choice and setting of protective and monitoring devices	Is ADS used as fault protection? The inspector must check that all circuits are adequately protected using protective devices of a suitable type and rating. Monitoring devices, such as residual current monitoring devices (RCMs) and insulation monitoring devices (IMDs), are less common.

(ii) Non-conducting location

Absence of protective conductors	This is very rarely found, although it may be present in specialised installations which are designed in accordance with Section 418.

(iii) Earth-free local equipotential bonding

Presence of earth-free local equipotential bonding	This is very rarely found, although it may be present in specialised installations which are designed in accordance with Section 418.

(iv) Electrical separation

Provided for one item of equipment	Is separation provided for one item? Does the transformer comply with the relevant standards?
	This section is common in many installations. An example could be a shaver point in a bathroom.
Provided for more than one item of equipment	Is separation provided for more than one item, such as where an isolating transformer supplies socket outlets in a laboratory or in a workshop where RCD protection is not suitable? It may also be used where regular work is required in a conducting location with restricted movement.
	This is less common but may apply in special circumstances.
	The requirements of Section 413 must be met.

Additional protection

Presence of RCDs	Are RCDs present and do they comply with Section 415.1?
	This is applicable in most installations where RCDs rated no more than 30 mA are used for additional protection, such as for socket outlets for general use and in mobile equipment that is used outdoors.
	If cables are concealed in the fabric of a building, the relevant box under the section 'Cables and conductors' applies.
Presence of supplementary bonding	Do we need to carry out supplementary bonding?
	Supplementary protective bonding conductors are installed to ensure that shock risk is substantially reduced under both earth fault and earth leakage conditions. It is essential to reduce the risk of touch voltage during fault or leakage conditions. See Regulations 415.2.1 and 415.2.2.
	Also see Regulation 411.3.2.6 for information regarding disconnection times.

Prevention of mutual detrimental influences

Proximity to non-electrical services	Are both electrical equipment and cables suitably distanced from any services that may affect the installation? An example may be hot water pipes near cables. If close spacing is unavoidable, the electrical parts must be suited for proximity to the services.
Segregation of Band I and Band II circuits	Are any electrical circuits operating at ELV adequately segregated from low-voltage circuits (unless the ELV circuits are insulated to low-voltage standards)? An example may be a door bell circuit using bell wire that must be segregated from lighting circuits. See Section 528.
Segregation of safety circuits	Are safety circuits such as fire-alarm circuits or centrally-fed emergency lighting completely segregated from all other circuits? These should never share containments, such as conduits. See Chapter 56.

Identification

Presence of diagrams, instructions, circuit charts	Does the electrical installation contain suitable diagrams and charts detailing specific information about the installation? See Regulation 514.9.1.
Presence of danger notices and other warning notices	Is there a range of danger and warning notices to satisfy the requirements of Section 514.9? *All* electrical installations require these.
Labelling of protective devices, switches and terminals	Are all terminals clearly marked? Do all switches and devices, such as fuses and circuit breakers, clearly identify their purpose? See Section 514.
Identification of conductors	Are all conductors clearly identified by colour or marking? This includes sleeving of conductors. The inspector must be satisfied that all conductors are in accordance with Section 514.

Cables and conductors

Selection of conductors for current-carrying capacity and voltage drop	Are all conductors suitably sized for the intended load and voltage drop constraints? The inspector will be reliant on the designer's specification for this, but a good level of experience is also required to make an informed judgement. The inspector must ensure correct coordination exists: $I_b \leq I_n \leq I_z$.
Erection methods	Are all electrical equipment, cables and containment systems suitably and securely installed? As an example, does a conduit have adequate saddles which are correctly spaced?
Routing of cables in prescribed zones	Are all cables that are concealed in the fabric of the building, within the 'zones of protection'? For example, do the concealed cables run vertically above or below an accessory as detailed in the IET On-Site Guide? Installations should be compliant with Section 522.6. This requirement works in conjunction with the two requirements that follow.
Cables incorporating earthed armour or sheath, or run within an earthed wiring system	Are cables present that are not in the zones of protection as detailed above? If so, they must be protected by an earthed metallic covering. Also, where cables are concealed in a wall and are not provided with additional protection by an RCD as below, they must comply with this section. Installations under effective supervision may not be applicable.
Additional protection provided by a 30 mA RCD for cables	Are cables concealed in a wall where the installation is not under effective supervision? If so, the applicable circuits must be provided with additional protection by an RCD in accordance with Section 415.1. If the requirements of the section above are satisfied and cables are suitably incorporated in an earthed metallic covering, this section is not applicable.
Connection of conductors	Are all connections secure and, where applicable, readily accessible for maintenance?
Presence of fire barriers, suitable seals and protection against thermal effects	Does all electrical equipment have suitable protection to stop the spread of fire and minimise thermal effects in accordance with Chapter 42? It must also be verified that elements of the building's structure that are affected by the installation, such as trunking that passes through floors, are adequately sealed.

General

Presence and correct location of appropriate devices for isolation and switching	Are all isolators and switches located correctly and suitable for their intended use? An example of this may be a switch for mechanical maintenance adjacent to a machine. The switch must be rated for on-load switching and be located in a suitable position in close proximity to the machine, as well as being accessible. See Chapter 53.
Adequacy of access to switchgear and other equipment	Is all switchgear and control equipment, as well as any accessories, fully accessible for use and maintainability? The particular requirements of Section 729 may also need to be met.
Particular protective measures for special installations and locations	Is this a special installation or location, for example a bathroom, swimming pool or caravan park? If the installation contains any locations or is an installation as detailed in Part 7 of BS 7671, the particular requirements must be satisfied. This is due to the additional risks associated with these special locations.
Connection of single-pole devices for protection or switching in line conductors only	The inspector must be satisfied that all single-pole devices control the line conductor of the circuit and not the neutral. This includes circuit breakers, fuses, switches etc.
Correct connection of accessories and equipment	Is the polarity of all equipment correct?
Presence of undervoltage protective devices	Could the loss of supply and subsequent restarting cause a danger? If so, is undervoltage protection provided? An example may be a machine controlled by a contactor such that, should the supply fail and be restored, the motor wouldn't automatically restart.

Selection of equipment and protective measures for external influences	Is all electrical equipment suitably selected and erected, taking into account potential external influences? A complete list of external influences can be seen in Appendix 5 of BS 7671.
Selection of appropriate functional switching devices	Does all current-using equipment have a suitably rated and located switch in order for the equipment to be used safely? This includes light switches, auxiliary circuits and motor control. The requirements of Section 537.5 must be met.

Information contained within a Schedule of Test Results (Form 4)

The completed Schedule of Test Results is a vital part of initial verification. It contains all the technical details which are used to verify that the installation is safe and suitable for use.

Each new circuit must be tested and all results need to be analysed to ensure that BS 7671 has been met.

The testing methods and processes have already been addressed at length in this publication.

This form must accompany the Electrical Installation Certificate.

THE CERTIFICATION PROCESS FOR A COMPLETED INSTALLATION

All documents must be completed and signed by a competent person or persons.

Assessment criteria

5.3 Describe the certification process for a completed installation and identify the responsibilities of different relevant personnel in relation to the completion of the certification process

Electrical Installation Certificate

The Electrical Installation Certificate, Schedule of Inspections and Schedule of Test Results must be handed over to the person ordering the work. All three documents must be complete and relevant to the work carried out.

Form 1: Short form of Electrical Installation Certificate

Form 1 applies when one person is responsible for the design, construction, inspection and testing of the electrical installation. All documents must be signed by the same person.

Form 2: Electrical Installation Certificate

Form 2 applies when there is more than one person involved in the process of certification.

Only authorised signatories may sign on behalf of the companies executing the design, construction, inspection and testing respectively. A signatory who is authorised to certify more than one category of work should sign in each of the appropriate places.

The designer must sign the first page of the certificate under 'Design'.

The installer must sign the first page of the certificate for the construction of the electrical installation.

The person inspecting and testing must sign as the person responsible for certifying the electrical installation. All Schedules of Inspection must also be signed.

Form 5: Minor Electrical Installation Works Certificate

The Minor Electrical Installation Works Certificate (or Minor Works Certificate) can *only* be used if the minor electrical work does not include the provision of a new circuit. A Minor Works Certificate indicates the responsibility for design, construction, inspection and testing of the work described on the certificate.

PROCEDURES AND REQUIREMENTS FOR DOCUMENTS AND CERTIFICATION

All electrical installation work must be designed, constructed, inspected and tested. When the inspection and testing is complete, the customer will be supplied with the relevant copies of the certification.

Electrical Installation Certificate

This safety certificate will be issued to confirm that the electrical installation work to which it relates has been designed, constructed, inspected and tested in accordance with British Standard 7671 (the IET Wiring Regulations). As part of the certification, the Schedule of Inspections and the Schedule of Test Results must be completed with satisfactory outcomes.

The recipient should receive an 'original' certificate and the contractor should retain a duplicate. If you are the person ordering the work, but not the owner of the installation, you should pass this certificate, or a full copy of it including the schedules, immediately to the owner.

The 'original' certificate should be retained in a safe place to be shown to any person inspecting or undertaking further work on the electrical installation in the future. If the property is later vacated, this certificate will demonstrate to the new owner that the electrical installation complied with the requirements of British Standard 7671 at the time the certificate was issued. The Construction (Design and Management) Regulations require that, for a project covered by those regulations, a copy of this certificate, together with schedules, is included in the project health and safety documentation.

Minor Electrical Installations Works Certificate

This certificate will be issued to confirm that the electrical installation work to which it relates has been designed, constructed, inspected and tested in accordance with British Standard 7671 (IET Wiring Regulations).

The recipient should receive an 'original' certificate and the contractor should retain a duplicate. If you are the person ordering the work, but not the owner of the installation, you should pass this certificate, or a copy of it, to the owner.

A separate certificate should be received for each existing circuit on which minor works have been carried out. This certificate is not appropriate if the contractor has undertaken more extensive installation work, for which an Electrical Installation Certificate is required.

The certificate should be retained in a safe place and be shown to any person inspecting or undertaking further work on the electrical installation in the future. If the property is later vacated, this certificate will demonstrate to the new owner that the minor electrical installation work carried out complied with the requirements of British Standard 7671 at the time the certificate was issued.

Schedule of Inspections

The Schedule of Inspections is completed and supplied as part of the electrical installation certificate. This document must be completed with no faults recorded and presented as part of the Electrical Installation Certificate.

Generic Schedule of Test Results

The Schedule of Test Results is completed and supplied as part of the Electrical Installation Certificate. This document must be completed with no faults recorded and presented as part of the Electrical Installation Certificate. Complex installations may require more than one Schedule of Test Results. One schedule must be completed for each distribution board and consumer control unit (including distribution circuits).

ASSESSMENT GUIDANCE

Be very careful when completing the Schedule of Inspections. An X or a blank are not allowed and consideration must be given to all of the installation.

KEY POINT

The original proof of the documentation must be given to the person ordering the work.

ASSESSMENT CHECKLIST

WHAT YOU NOW KNOW/CAN DO

Learning outcome	Assessment criteria	Page number
1 Understand the principles, regulatory requirements and procedures for completing the safe isolation of an electrical circuit and complete electrical installations in preparation for inspection, testing and commissioning	*The learner can:* **1** State the requirements of the Electricity at Work Regulations 1989 for the safe inspection of electrical systems and equipment, in terms of those carrying out the work and those using the building during the inspection	2
	2 Specify and undertake the correct procedure for completing safe isolation	4
	3 State the implications of carrying out safe isolations to: ■ other personnel ■ customers/clients ■ public ■ building systems (loss of supply)	7
	4 State the implications of not carrying out safe isolations to ■ self ■ other personnel ■ customers/clients ■ public ■ building systems (presence of supply)	8
	5 Identify all health and safety requirements which apply when inspecting, testing and commissioning electrical installations and circuits including those which cover: ■ working in accordance with risk assessments/permits to work/method statements ■ safe use of tools and equipment ■ safe and correct use of measuring instruments ■ provision and use of PPE ■ reporting of unsafe situations.	10

Learning outcome	Assessment criteria	Page number
2 Understand the principles and regulatory requirements for inspecting, testing and commissioning electrical systems, equipment and components	*The learner can:*	
	1 State the purpose of and requirements for initial verification and periodic inspection of electrical installations	13
	2 Identify and interpret the requirements of the relevant documents associated with the inspection, testing and commissioning of an electrical installation	13
	3 Specify the information that is required to correctly conduct the initial verification of an electrical installation in accordance with the IET Wiring Regulations and IET Guidance Note 3.	24
3 Understand the regulatory requirements and procedures for completing the inspection of electrical installations	*The learner can:*	
	1 Identify the items to be checked during the inspection process for given electrotechnical systems and equipment, and their locations as detailed in the IET Wiring Regulations	32
	2 State how human senses (sight, touch, etc) can be used during the inspection process	27
	3 State the items of an electrical installation that should be inspected in accordance with IET Guidance Note 3	28

	4 Specify the requirements for the inspection of the following:
	■ earthing conductors
	■ circuit protective conductors
	■ protective bonding conductors:
	   □ main bonding conductors
	   □ supplementary bonding conductors
	■ isolation
	■ type and rating of overcurrent protective devices.
		38

Learning outcome	Assessment criteria	Page number
4 Understand the regulatory requirements and procedures for the safe testing and commissioning of electrical installations	*The learner can:*	
	1 State the tests to be carried out on an electrical installation in accordance with the IET Wiring Regulations and IET Guidance Note 3	43
	2 Identify the correct instrument for the test to be carried out in terms of: ■ instrument's fitness for purpose ■ identifying the right scale/settings of the instrument appropriate to the test being carried out	45
	3 Specify the requirements for the safe and correct use of instruments to be used for testing and commissioning	49
	4 Explain why it is necessary for test results to comply with standard values and state the actions to take in the event of unsatisfactory results being obtained	51
	5 Explain why testing is carried out in the exact order as specified in the IET Wiring Regulations and IET Guidance Note 3	52
	6 State the reasons why it is necessary to verify the continuity of circuit protective conductors, earthing conductors, bonding conductors and ring final circuit conductors	52
	7 Specify and apply the methods for verifying the continuity of circuit protective conductors and ring final circuit conductors and interpreting the obtained results	52
	8 State the effects that cables connected in parallel and variations in cable length can have on insulation resistance values	65
	9 Interpret and apply the procedures for completing insulation resistance testing	65
	10 Explain why it is necessary to verify polarity	73
	11 Interpret and apply the procedures for testing to identify correct polarity	73

Learning outcome	Assessment criteria	Page number
	12 Specify and apply the methods for measuring earth electrode resistance and correctly interpreting the results	81
	13 Identify the earth fault loop paths for the following systems: ■ TN-S ■ TN-C-S ■ TT	75
	14 State the methods for verifying protection by automatic disconnection of the supply	87
	15 Specify the methods for determining prospective fault current	99
	16 Specify the methods for testing the correct operation of residual current devices (RCDs)	103
	17 State the methods used to check for the correct phase sequence	107
	18 Explain why having the correct phase sequence is important	107
	19 State the need for functional testing and identify items which need to be checked	108
	20 Specify the methods used for verification of voltage drop	110
	21 State the cause of voltage drop in an electrical installation	110
	22 State the appropriate procedures for dealing with customers and clients during the commissioning and certification process.	113

Learning outcome	Assessment criteria	Page number
5 Understand the procedures and requirements for the completion of electrical installation certificates and related documentation	*The learner can:*	
	1 Explain the purpose of and relationship between: ■ an electrical installation certificate ■ a minor electrical installation works certificate ■ a schedule of inspections ■ a schedule of test results	115
	2 State the information that must be contained within: ■ an electrical installation certificate ■ a minor electrical installation works certificate ■ a schedule of inspections ■ a schedule of test results	122
	3 Describe the certification process for a completed installation and identify the responsibilities of different relevant personnel in relation to the completion of the certification process	131
	4 Explain the procedures and requirements, in accordance with the IET Wiring Regulations, IET Guidance Note 3 and where appropriate customer/client requirements for the recording and retention of completed: ■ electrical installation certificates ■ minor electrical installation works certificates ■ schedule of inspections ■ schedule of test results.	132

ASSESSMENT GUIDANCE

This unit is assessed in the following ways:

- Closed book online e-volve multiple-choice assessment
- Practical assessment
- Closed book written examination
- Common task – safe isolation. This is common to 2357 Units 305, 306, 307 and 308.

The assessments used in 2357 Unit 307 are in line with the assessments used for the City & Guilds 2394 Certificate in Initial Verification of Electrical Installations. Once you successfully complete the 2357 NVQ, you will also be awarded a certificate for the 2394 qualification.

Online multiple-choice assessment

- This is a closed book online e-volve multiple-choice assessment.
- You are allowed to take with you a copy of the IET On-Site Guide.
- A calculator (non-programmable) may be required for some of the questions.
- Attempt all questions.
- Do not leave until you are confident that you have completed all questions.
- Keep an eye on the time, as it moves quickly when you are concentrating.
- Make sure you read each question fully before answering.
- Ensure you know how the e-volve system works. Ask for a demonstration if you are not sure.
- Do not take any paperwork into the exam with you.
- If you need paper to work anything out, ask the invigilator to provide some.
- Make sure your mobile phone is switched off (not on silent) during the exam. You may be asked to give it to the invigilator.

Before the assessment

- You will find some questions starting on page 141 to test your knowledge of the learning outcomes.
- Make sure you go over these questions in your own time.
- Spend time on revision in the run-up to the assessment.

Practical assessment

- The practical assessment requires you to carry out electrical installation testing on a rig.
- Make sure you know the sequence of testing.
- You may have guidance material with you.
- You must demonstrate competence to your assessor.
- Ensure you check all instruments before use, to ensure they work correctly and are in a safe condition.
- Keep rechecking your instruments for accuracy when carrying out continuity tests; changing leads will necessitate re-nulling or zeroing.
- Ensure you use all necessary Personal Protection Equipment; this should be made available to you.
- Testing is not a race. Doing it right is far better than doing it quickly.
- When you complete a particular stage or test, ensure everything is reconnected or put back before moving on.
- Do not create clutter in your work area. Work tidily and things become easier.
- Ask before you switch off: ask before you switch on.
- Make sure you fill in any paperwork fully. Do not leave any gaps.

Above all else – WORK SAFELY.

Written examination

- For the written examination you will answer six questions. Three questions will be scenario-based and three questions will be general testing questions. The time allowed is 1 hour 30 minutes.
- This is a closed book examination – you are not allowed to take any textbooks or guidance into the exam room with you.
- You will need writing and drawing materials and a scientific non-programmable calculator.
- Make sure you arrive on time for any assessment you take.
- If you are unsure of anything, ask – be clear what you are doing.

Common task

- This task is common to 2357 Units 305, 306, 307 and 308. You only need to carry out the task once.
- You will be required to answer a number of short answer questions for which you may research the answers. These questions relate to safe isolation. You will then be required to perform a safe isolation procedure under the guidance of your assessor.

OUTCOME KNOWLEDGE CHECK

1   Which one of the following is a statutory document?

a)   BS 7671 IET Wiring Regulations 2008.

b)   Electricity at Work Regulations 1989.

c)   IET On-Site Guide.

d)   IET Guidance Notes 3.

2   Test leads used with instruments operating above 50 V a.c. should comply with:

a)   BS 7671 IET Wiring Regulations 2008

b)   GS 38

c)   BS 88

d)   BS EN 61215 2005.

3   The measured values for L-L, N-N and cpc-cpc for a ring final circuit test are L-L 0.8 Ω, N-N 0.8 Ω and cpc-cpc 1.4 Ω. The expected values when testing L-cpc at each outlet would be:

a)   0.55 Ω

b)   0.75 Ω

c)   2.2 Ω

d)   3 Ω.

4   The instrument used to carry out a continuity test on a bonding conductor would be:

a)   an earth electrode tester

b)   an earth loop impedance tester

c)   a continuity tester

d)   a low resistance ohmmeter.

5   An insulation resistance test is carried out on a four-way consumer unit. Each circuit is tested individually with the following results.

Circuit	One	Two	Three	Four
Test value (MΩ)	20	100	50	5

The overall test result would be:

a)   3.57 MΩ

b)   4 MΩ

c)   5 MΩ

d)   175 MΩ.

6   The document to be completed following the addition of a light to an existing circuit would be which of the following?

a)   Electrical Installation Certificate.

b)   Electrical Installation Condition Report.

c)   Minor Electrical Installation Works Certificate.

d)   Electrical Installation Report.

7   Which human sense would be used to verify correct identification of conductors?

a)   Taste.

b)   Smell.

c)   Touch.

d)   Sight.

8   An earth loop impedance test is carried out at a socket outlet. The value obtained is known as:

a) $R_1 + R_2$

b) Z_s

c) Z_e

d) $Z_s + (R_1 + R_2)$.

9   The maximum allowable voltage drop in a circuit is 6.9 V. The cable length is 15 m and the voltage drop 22 MV/A/m. The maximum load current will be:

a)   15 A

b)   20.9 A

c)   22 A

d)   22.8 A.

10   An RCD tester displays the test result in:

a)   amperes

b)   mA

c)   seconds

d)   milliseconds.

KNOWLEDGE CHECK

In your written examination there will be three scenario-based questions and three general testing questions. The questions below reflect the general testing questions.

1   A ring final circuit test is to be carried out on a new installation wired in 2.5 mm² single core PVC cables in PVC conduit. Each loop is 60 metres long and the resistance of the cable is 7.41 milliohms per metre.

   a) Describe the three stages of the ring final circuit test.

   b) State the instrument to be used.

   c) Calculate the resistance of each loop.

   d) Calculate the expected value at each outlet.

   e) Describe a method of identifying the ring conductors' legs at the board.

2   A new installation is to be tested and certified.

   a) Identify the three forms/certificates to be completed.

   b) State the three persons responsible for signing the form(s).

   c) State the title of the document used for the initial verification of additions or alterations to existing circuits.

   d) The test voltage and minimum acceptable insulation resistance value for a 400 V a.c motor circuit.

3   A domestic dwelling is supplied by a TN-C-S system. A new storage shed clad in corrugated steel sheet is to be erected 15 m from the property. The electrical installation within the shed will be wired using single-core cable in galvanised steel conduit supplying a number of 13 A socket-outlets and fluorescent luminaires. State:

   a) the system to be used for the shed supply

   b) the reason for your choice

   c) what device would be used to provide additional protection

   d) the rating of the device in c) and the maximum permitted disconnection time when tested at 5 times the rating

   e) the method of providing an earth connection.

UNIT 308

Understanding the principles, practices and legislation for diagnosing and correcting electrical faults in electrotechnical systems and equipment in buildings, structures and the environment

There are many types of fault that can occur in electrical circuits. These include not only the failure of physical components but also errors that may have been introduced during the original design or during the installation. This unit is designed to enable learners to understand principles, practices and legislation associated with diagnosing and correcting electrical faults in electrotechnical systems and equipment in buildings, structures and the environment in accordance with statutory and non-statutory regulations and requirements. Its content is the knowledge needed by a learner to underpin the application of skills used for fault diagnosis and correction in electrotechnical systems and equipment in buildings, structures and the environment.

LEARNING OUTCOMES

There are five learning outcomes to this unit. The learner will:

1 understand the principles, regulatory requirements and procedures for completing the safe isolation of electrical circuits and complete electrical installations

2 understand how to complete the reporting and recording of electrical fault diagnosis and correction work

3 understand how to complete the preparatory work prior to fault diagnosis and correction work

4 understand the procedures and techniques for diagnosing electrical faults

5 understand the procedures and techniques for correcting electrical faults.

Within this book, the assessment criteria within the learning outcomes will, on occasions, be re-ordered. This is to ensure good continuity of the subject matter.

This unit will be assessed by:

■ project-based assignments

■ practical assessment.

OUTCOME 1

Understand the principles, regulatory requirements and procedures for completing the safe isolation of electrical circuits and complete electrical installations

Assessment criteria

1.1 Specify and undertake the correct procedure for completing the safe isolation of an electrical circuit

Duty holder

The person in control of the danger is the duty holder. This person must be competent by formal training and experience and with sufficient knowledge to avoid danger. The level of competence will differ for different items of work.

HEALTH AND SAFETY LEGISLATION

The basic concept of 'health and safety legislation' is to provide the legal framework for the protection of people from illness and physical injury that may occur in the workplace.

The HSW Act is the basis of all British health and safety law. It provides a comprehensive and integrated piece of legislation that sets out the general duties that employers have towards employees, contractors and members of the public, and that employees have to themselves and to each other. These duties are qualified in the HSW Act by the principle of 'so far as is reasonably practicable'.

What the law expects is what good management and common sense would lead employers to do anyway: that is, to look at what the risks are and take sensible measures to tackle those risks. The person(s) who are responsible for the risk and best placed to control that risk are usually designated the **duty holder**.

The HSW Act, which is an enabling act, is based on the principle that those who create risks to employees or others in the course of carrying out work activities are responsible for controlling those risks. The Act places specific responsibilities on the following:

- employers
- the self-employed
- employees
- designers
- manufacturers and suppliers
- importers.

The HSW Act lays down the general legal framework for health and safety in the workplace with specific duties being contained in Regulations also called Statutory Instruments (SIs) which are also examples of laws approved by Parliament.

The HSW Act, and general duties in the Management of Health and Safety at Work Regulations 1999, are goal setting and leave employers the freedom to decide how to control risks which they identify.

However some risks are so great or the proper control measures so costly, that it would not be appropriate to leave employers to decide what to do about them. Regulations identify these risks and set out the specific action that must be taken.

The Electricity at Work Regulations 1989 (EAW) are the main statutory regulations that deal with work on electrical systems and require people in control of electrical systems to ensure they are safe to use and maintained in a safe condition.

GUIDANCE AND NON-STATUTORY REGULATIONS

The HSE and other organisations publish guidance and non-statutory regulations on a range of subjects. Guidance can be specific to the health and safety problems of an industry or of a particular process used in a number of industries.

The main purposes of guidance are:

- to interpret and help people to understand what the law says
- to help people comply with the law
- to give technical advice

Following guidance and non-statutory regulations is not compulsory and employers are free to take other action, but if they do follow the guidance they will normally be doing enough to comply with the law.

One very good example of guidance and non-statutory regulations is BS 7671:2008 Requirements for Electrical Installations (the IET Wiring Regulations 17th edition), more usually known as the 'IET Wiring Regs'. If electrotechnical work is undertaken in accordance with BS 7671 it is almost certain to meet the requirements of the statutory regulations, the Electricity at Work Regulations, dealing with work with electrical equipment and systems.

BS 7671 is the national standard in the United Kingdom for low voltage electrical installations. The document is largely based on the European Committee for Electrotechnical Standardisation (CENELEC) harmonised documents, and is therefore technically very similar to the current wiring regulations of other European countries. The regulations deal with the design, selection, erection, inspection and testing of electrical installations operating at a voltage up to 1000 V a.c.

Electricity

BS 7671, which are non-statutory regulations, relate principally to the design and erection of electrical installations so as to provide for safety and proper functioning for the intended use. If an installation is constructed in accordance with BS 7671, protection should be afforded to the user from:

■ electric shock

■ excessive temperature

■ under-voltage, overvoltage and electromagnetic disturbances

■ power supply interruptions

■ arcing and burning.

However, a great number of electrical accidents occur because people are working on or near equipment that is thought to be dead but which is in fact live, or the equipment is known to be live but those involved in the work do not have adequate training or appropriate equipment, or they have not taken adequate precautions.

EAW Regulation 4(3) requires that every work activity, including operation, use and maintenance of a system and work near a system, shall be carried out in such a manner as not to give rise, so far as is reasonably practicable, to danger.

The EAW Regulations provide for two basic types of system of work: work on de-energised conductors (Regulation 13) and work on live conductors (Regulation 14). Regulation 13 covers the preferred system of working, which is to remove the danger at source, ie by making the conductors dead. Regulation 14 requires adequate safeguards to protect the person at work from the hazards of live conductors.

Many accidents to electricians, technicians and electrical engineers occur when they are working on equipment that could have been isolated. In most cases, adequate planning and work programming will allow such jobs to be carried out as the Regulations require, ie with the equipment dead and **isolated**.

SAFE ISOLATIONS

Isolation and switching of supply

The preparation of electrical equipment for fault diagnosis and repair purposes often requires effective disconnection from all live supplies and a means for securing that disconnection, for example, by locking off with a suitable lock and single key. There is an important distinction between switching and isolation.

Isolation

Isolation is the disconnection and separation of the electrical equipment from every source of electrical energy in such a way that this disconnection and separation is secure.

Assessment criteria

1.2 State the implications of carrying out safe isolations to:

■ other personnel
■ customers/clients
■ public
■ building systems (loss of supply)

Switching is cutting off the supply while isolation is the secure disconnection and separation from all sources of electrical energy. A variety of control devices are available for switching, isolation or a combination of these functions, some incorporating protective devices. Before starting work on a piece of isolated equipment, checks should be made, using an approved testing device, to ensure that the circuit is dead. See below for details of a standard isolation procedure.

There are also secondary hazards associated with electrical testing that must be considered when risk assessments and a safe work practice document are being prepared.

A standard procedure for isolation is as follows.

1   Select an approved voltage indicator to GS 38 and confirm operation.

2   Locate correct source of supply to the section needing isolation.

3   Confirm that the device used for isolation is suitable and may be secured effectively.

4   Power down circuit loads if the isolator is not suitable for on-load switching.

5   Disconnect using the located isolator (from step 2).

6   Secure in the off position, keep key* on person and post warning signs.

7   Using voltage indicator, confirm isolation by checking ALL combinations.

8   Prove voltage indicator on known source, such as proving unit.

*If the device used for the purpose of isolation is a fuse or removable handle instead of a lockable device, keep this securely under supervision while work is undertaken.

Carrying out a safe isolation procedure will safeguard not only the person undertaking work on the installation, but also other people who may be working within the building, such as:

■   occupiers and other trades people (who may be inconvenienced by loss of supply to essential equipment/machinery)

■   the customer or client for whom the work is being done (who may suffer from loss of service and downtime)

■   members of the public (who are exposed to possible danger due to loss of essential services such as fire-alarm systems, emergency or escape lighting)

■   those who require the continued provision of a supply for data and communication systems.

Remember that, if isolation is required to be carried out on a distribution board that is in a communal area, such as at the entrance to flats, a shop or hotel, there will be additional requirements such as barriers and notices to prevent unauthorised access to the work area.

Most electrical accidents occur because people are working on or near equipment that is:

- thought to be dead but which is live
- known to be live, but those involved do not have adequate training or appropriate equipment, or they have not taken adequate precautions.

CONSEQUENCES OF NOT ENSURING SAFE ISOLATIONS

If the correct isolation procedure is *not* undertaken, it can have implications not only for the person who is carrying out the work on the installation but to other people, such as the occupiers of the building, clients or customers, other trades or personnel working within the building and members of the public.

The implications:

- to the person carrying out the work and to members of the public are the possibility of electric shock or burns
- to the occupiers of the building include a risk of contact with electrical parts when basic protection has been removed
- to clients and customers will be risk of shock or burns and damage to equipment.

An incorrect isolation procedure may also present a risk of damage to electrical equipment and to building fabric.

SAFE WORKING PROCEDURES FOR FAULT DIAGNOSIS

Electricity is a safe, clean and powerful source of energy and is in use in practically every factory, office, workshop and home in the country. However, this energy source can also be very hazardous – with a risk of causing death, if it is not treated with care. Injury can occur when live electrical parts are exposed and can be touched, or when metalwork that is meant to be earthed becomes live at a dangerous voltage.

The possibility of touching live parts is increased during electrical testing and fault-finding, when conductors at dangerous voltages are often exposed. This risk can be reduced if testing is done while the equipment is isolated from any source of electrical supply. However, this is not always possible and, if this is the case, it is important to follow procedures that prevent contact with any hazardous voltages internal to the system.

Assessment criteria

1.3 State the implications of not carrying out safe isolations to:

- self
- other personnel
- customers/clients
- public
- building systems (presence of supply)

ASSESSMENT GUIDANCE

Never take any action that may put yourself or anyone else in danger. When undertaking the unit assessment you must work safely at all times.

Assessment criteria

1.4 Identify all health and safety requirements which apply when diagnosing and correcting electrical faults in electrotechnical systems and equipment including those which cover:

- working in accordance with risk assessments/permits to work/method statements
- safe use of tools and equipment
- safe and correct use of measuring instruments
- provision and use of PPE
- reporting of unsafe situations.

ACTIVITY

Identify four faults that could occur from the input of a 20 A triple-pole isolator through a direct online starter to a three-phase motor.

But before fault finding and testing work commences, reference must be made to the following relevant legal duties.

SmartScreen Unit 308
Handout 9 and Worksheet 9

■ *The Electricity at Work Regulations 1989* (EAWR): this is the principal legislation relating to work on electrical systems. Regulation 4(3) requires that: 'work on or near to an electrical system shall be carried out in such a manner as not to give rise, so far as is reasonably practicable, to danger'. The regulations provide for two basic types of system of work: work on de-energised conductors (Regulation 13) and work on live conductors (Regulation 14).

■ *The Management of Health and Safety at Work Regulations 1999*: these require employers to assess the risks to the health and safety of their employees while they are at work, in order to identify and put in place the necessary precautions to ensure safety. Depending on the extent of the work, this requires either a formal written risk assessment or one that relies on a generic system for simple work activities. A sample risk assessment is shown below.

> **KEY POINT**
>
> Regulation 13 covers the preferred system of working, which is to remove the danger at source by making the conductors dead.
>
> Regulation 14 requires adequate safeguards to protect the person at work from the hazards of live conductors.

What are the hazards?	Who might be harmed and how?	What actions have been taken?	Any further actions required to manage this risk?	Action by whom?	Action by when?	Done
Electric shock	Staff carrying out fault finding. Failure to carry out correct isolation procedure prior to testing; Working on equipment that is live; Failure to provide suitable barriers where live equipment needs to be shrouded; Use of faulty test equipment; Failure to use appropriate PPE Unsafe situations left unresolved. Visitors to and occupants of the building. Test area not properly guarded and notices not posted; Making contact with exposed conductive parts during testing	All electricians have received training in the following: Correct method of isolation; Safe testing procedures; Safe use of test instruments First aid training; Correct selection and use of PPE Correct procedure regarding action to be taken if unsafe situations are identified All electricians have received training in the following: Safe work area demarcation – barriers, notices.	Regular audits by supervisor to confirm correct testing procedures are being observed; Refresher training for all electricians.	Supervisor	01/02/2013	01/02/2013
Lone working	Lone worker. Increased risk of injury due to being unable to summon assistance.	All electricians have received effective means of communication (eg landline or mobile phone). Any person working alone will notify their supervisor of their itinerary including where they will be working and what time they expect to finish.	No	Supervisor	01/02/2013	01/02/2013
Asbestos	Staff carrying out fault finding. Failure to identify presence of asbestos and drilling or abrading asbestos board or ceiling tiles	All electricians have received training in the following: Asbestos awareness A risk assessment has been carried out and a safe system of work prepared.				
Slips trips and falls	Staff and visitors. May be injured if they trip over objects or slip on spillages	General good housekeeping. All areas are well lit including stairs. There are no trailing leads or cables. Staff to keep work areas clear, eg no boxes left in walkways.	Better housekeeping is needed in staff kitchen, eg on spills.	Site responsible person. All staff, supervisor to monitor	01/02/2013	01/02/2013

Sample risk assessment

The method statement

Once the risk assessment has been carried out, a safe system of work, sometimes called a method statement, can be prepared to enable the work to be undertaken in a safe manner. Risk assessments and method statements are often known collectively by the acronym RAMS: **R**isk **A**ssessment and **M**ethod **S**tatement.

The safe system of work or method statement will include such items as:

- who is authorised to undertake testing and, where appropriate, how to access a test area and who should not enter the test area
- arrangements for isolating equipment and how the isolation is secured
- provision and use of personal protective equipment (PPE) where necessary
- the correct use of tools and equipment
- the correct use of additional protection measures, for example, flexible insulation that may need to be applied to the equipment under test while its covers are removed
- use of barriers and positioning of notices
- safe and correct use of measuring instruments
- safe working arrangements to be agreed with client, duty holder or responsible person
- how defects are to be reported and recorded
- instructions regarding action to be taken if unsafe situations are identified.

Details of safe systems of work or safe working procedures for fault diagnosis, testing and fault repair activities should, wherever it is reasonably practicable to do so, be written down.

Guidance on live and dead working can be found in the HSE publication *Memorandum of guidance on the Electricity at Work Regulations 1989* (HSR 25) and this can be downloaded free of charge from the Health and Safety Executive website: www.hse.gov.uk

Safe working procedures should be reviewed regularly, to make sure that they are being followed and are still appropriate for the work that is being carried out. If any changes are made to the procedures, all people who are involved in the fault diagnosis regime should be given relevant instruction and training.

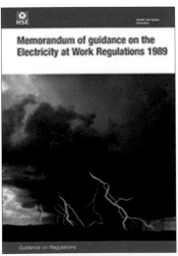

Refer to this publication for advice on live and dead working

Permit to work

A permit to work (PTW) procedure is a specialised written safe system of work that ensures potentially dangerous work, such as work on high-voltage electrical systems (above 1000 V) or complex lower-voltage electrical systems, is done safely. The PTW also serves as a means of communication between those controlling the danger (the duty holder) and those who carry out the hazardous work. Essential features of PTW systems are:

- clear identification of who may authorise particular jobs (and any limits to their authority), and who is responsible for specifying the necessary precautions

- that the permit should only be issued by a technically competent person, who is familiar with the system and equipment, and who is authorised in writing by the employer to issue such documents

- provision of training and instruction in the issue, use and closure of permits

- monitoring and auditing to ensure that the PTW system works as intended

- clear identification of the types of work considered hazardous

- clear and standardised identification of tasks, risk assessments, permitted task duration and any additional activity or control measure that occurs at the same time.

A permit to work should state clearly:

- the person the permit is addressed to, that is the person carrying out the work or the leader of the group or working party who will be present throughout the work

- the exact equipment that has been made dead and its precise location

- the points of isolation

- where the conductors are earthed (on high-voltage systems)

- where warning notices are posted and special safety locks fitted

- the nature of the work to be carried out

- the presence of any other source of hazard, with cross-reference to other relevant permits

- further precautions to be taken during the course of the work.

The effective operation of a PTW system requires involvement and cooperation from a number of people, and the procedure for issuing a PTW should be written down and adhered to.

> **ASSESSMENT GUIDANCE**
>
> Write clearly the instructions on the permit to work so there is no chance of misunderstanding what is required.

Planning and agreeing procedures

Before fault diagnosis is carried out, the safe-working arrangements must be discussed and agreed with the client and/or the duty holder (or responsible person for the installation) in a clear, concise and courteous manner. The testing and inspection procedure must be a planned activity as it will almost certainly affect people who work or live in the premises where the installation is being tested. This ensures that everyone who is concerned with the work understands what actions need to be taken, such as:

- which areas of the installation may be subject to disconnection
- anticipated disruption times
- who might be affected by the work
- health and safety requirements for the site
- which area will have restricted access
- whether temporary supplies will be required whilst the fault diagnosis is underway
- reaching agreement on who has authority for the diagnosis and repair.

It may be that a specific person has responsibility for the safe isolation of a particular section of an installation and that person should be identified and the isolation arrangements agreed. By entering into dialogue with the client before work commences, the potential for unforeseen events will be minimised and good customer relations will be fostered.

For example, in an office block where the electrical installation is complex and provides supplies to many different tenants located on a number of floors, the safe isolation of a sub-circuit for testing purposes may require a larger portion of the installation to be turned off initially. In order to achieve this with minimal disruption, an agreement must be reached between the competent person tasked to carry out the work and the person responsible for the installations affected. This responsible person could be the office manager, the designated electrical engineer for the site or, in some cases, the landlord of the building.

Everyone involved in the work (for example, client, electrician and those in the workplace) has a responsibility for their own health and safety and that of others who may be affected by the work. Communication between all parties will ensure compliance with the respective health and safety requirements.

KEY POINT

You should appreciate the difference between the duty holder and the responsible person.

The person in control of the danger is the *duty holder*. This person must be competent by formal training and experience and with sufficient knowledge to avoid danger. The level of competence will differ for different items of work.

The person who is designated the *responsible person* has delegated responsibility for certain aspects of a company's operational functions such as fire safety, electrical operational safety or the day-to-day responsibility for controlling any identified risk such as *Legionella* bacteria.

ACTIVITY

Using any source of reference material, obtain a copy of a permit to work and determine when a PTW might be used on a low voltage installation.

SAFE AND CORRECT USE OF MEASURING INSTRUMENTS

In fault diagnosis, the use of suitable and safe voltage-indicating devices and measuring instruments is as important as the competency of the person undertaking the fault-finding activities. The possibility of touching live parts is increased during electrical testing and fault finding, when conductors at dangerous voltages are often exposed. The risks can be reduced if testing is done while the equipment or part of an installation is made dead and is isolated from any dangerous source of electrical supply. Special attention should be paid when carrying out tests with instruments capable of generating test voltages greater than 50 V or which use the supply voltage for the purpose of earth-loop testing or a residual-current device test. Refer to pages 187–188 for more information about electric shock.

Use of instruments for fault diagnosis

The HSE has produced Guidance Note GS 38 (Electrical test equipment for use by electricians), which is intended to provide guidance for electrically competent people, including electricians, electrical contractors, test supervisors, technicians, managers and/or appliance retailers. It offers advice in the selection and use of test probes, leads, lamps, and voltage-indicating devices and measuring equipment for circuits with rated voltages not exceeding 650 V. It recommends the use of fused test leads aimed primarily at reducing the risks associated with arcing under fault conditions. Where possible, it is recommended that tests are carried out at reduced voltages, which will usually reduce the risk of injury.

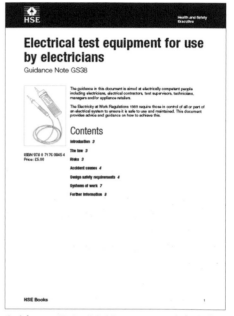

Guidance Note GS 38, an essential tool for electricians and others

Probes and leads

Guidance Note GS 38 recommends that probes and leads used in conjunction with voltmeters, multi-meters and electricians' test lamps or voltage indicators should be selected to prevent danger.

Probes should:

- have finger barriers or stops, or be shaped so that the hand or fingers cannot make contact with live conductors under test
- be insulated to leave an exposed metal tip not exceeding 4 mm measured across any surface of the tip. Where practicable, it is strongly recommended that this is reduced to 2 mm or less, or that spring-loaded retractable screened probes are used.

Leads should:

- be adequately insulated and coloured so that any one lead is readily identifiable from any other
- be flexible and sufficiently robust

- be long enough for the purpose – but not too long

- not have accessible exposed conductors, even if they become detached from the probe or from the instrument

- have suitably high breaking capacity (hbc), sometimes known as hrc, fuse or fuses with a low current rating (usually not exceeding 500 mA), or a current-limiting resistor and a fuse.

GS 38 also recommends that, if a test for the presence or absence of voltage is being made, the preferred instrument to be used is a proprietary test lamp or voltage indicator. The use of multi-meters for voltage indication has often resulted in accidents due to the multi-meter being set on the incorrect range.

Using instruments safely

When using test instruments to carry out fault diagnosis, follow these basic precautions to achieve safe working.

- *Understanding the equipment* – make sure you are familiar with the instrument to be used and its ranges; check its suitability for the characteristics of the installation it will be used on.

- *Self test* – many organisations regularly test instruments on known values to ensure they remain accurate. These tests would be documented.

- *Calibration* – all electrical test instruments should be calibrated on a regular basis. The time between calibrations will depend on the amount of usage that the instrument receives, although this should not exceed 12 months under any circumstances. Instruments have to be calibrated under laboratory conditions, against standards that can be traced back to national standards; therefore, this usually means returning the instrument to a specialist laboratory. Once calibrated, the instrument will have a calibration label attached to it stating the date the calibration took place and the date the next calibration is due. It will also be issued with a calibration certificate, detailing the tests that have been carried out, and a reference to the equipment used. Instruments that are subject to any electrical or mechanical misuse (for example, if the instrument undergoes an electrical short circuit or is dropped) should be returned for recalibration before being used again. Electrical test instruments are relatively delicate and expensive items of equipment and should be handled with care. When not in use, they should be stored in clean, dry conditions at normal room temperature. Care should also be taken of instrument leads and probes, to prevent damage to their insulation and to maintain them in safe working condition.

- *Check test leads* – make sure that these and any associated probes or clips are in good order, are clean and have no cracked or broken insulation. Where appropriate, the requirements of the HSE Guidance Note GS 38 should be observed for test leads.

- *Select appropriate scales and settings* – it is essential that the correct scale and settings are selected for an instrument. Manufacturers' instructions must be observed under all circumstances.

Tools and equipment

The responsibilities of users of work equipment are covered in Section 7 of the HSW Act and Regulation 14 of the Management Regulations. Work equipment includes hand tools and hand-held power tools.

Section 7 of the HSW Act requires employees to take reasonable care for themselves and for others who may be affected by their acts or omissions at work and to cooperate with the employer to enable the employer to discharge his duties under the Act.

Regulation 14 of the Management Regulations requires employees to use equipment properly in accordance with instructions and training and to report to the employer any dangerous situations that might arise during the course of their work.

The hand tools that are most commonly used in electrical installation work are: screwdrivers, pliers, side cutters, hacksaw, adjustable spanners, hammers, chisels, cold chisels, files, centre punches, wire strippers, cable strippers, crimpers, ratchet hand, scissors, etc.

Side cutters

Bolster chisel and lump hammer

Insulated screwdriver (courtesy of Insulated Tools Ltd)

The most common injuries from the use of hand tools are:

- blows and cuts to the hands or other parts of the body
- eye injuries due to the projection of fragments or particles
- sprains due to very abrupt movements or strains
- contact with live conductors.

The principal causes of injury are:

- inappropriate use of the tools
- use of faulty or inappropriate tools
- use of poor quality tools
- not using personal protection equipment
- forced postures.

Preventive measures:

- use quality tools in accordance with the type of work being carried out
- ensure employees are trained in the correct use of tools
- use approved insulated hand tools (IEC 60900) when working in the vicinity of live parts
- use protective goggles or glasses (BS EN 166) in all cases and above all when there is a risk of projected particles
- use gloves to handle sharp tools
- periodically check tools (repair, sharpening, cleaning, etc)
- periodically check the state of handles, insulating coverings, etc
- store and/or transport tools in boxes, tool bags or on suitable panels, where each tool has its place.

Hand-held power tools (portable electrical equipment)

Almost 25% of all reportable accidents involve portable electrical equipment, the majority being caused by electric shock. Electrical equipment that is hand held or being handled when switched on presents a large degree of risk to the user if it does develop a fault.

The principal causes of accidents are:

- use of damaged, defective or unsuitable equipment
- lack of effective maintenance
- misuse of equipment
- 'unauthorised' equipment (eg electric heaters, kettles, coffee percolators, electric fans) being used by employees
- use of power tools in a harsh environment (construction sites, etc).

Damaged transformer and poorly maintained equipment lead to a high risk of injury

Preventive measures:

- correct selection and use of equipment
- ensure employees are trained in the correct use of tools
- effective maintenance regime for all power tools, including user checks
- keeping of test records following portable appliance testing
- removal of damaged equipment from use.

PERSONAL PROTECTIVE EQUIPMENT AND ITS USE

Virtually all **personal protective equipment (PPE)** is covered by the Personal Protective Equipment at Work Regulations 1992 (PPE Regulations). The exception is respiratory equipment, which is covered by specific regulations relating to specific substances (lead, gases, substances hazardous to health, etc).

PPE is defined in the PPE Regulations as 'all equipment (including clothing affording protection against the weather) which is intended to be worn or held by a person at work and which protects them against one or more risks to their health or safety'. Such equipment includes safety helmets, gloves, eye protection, high-visibility clothing, safety footwear and safety harnesses. Employers are responsible for providing, replacing and paying for PPE.

Hearing protection and respiratory protective equipment provided for most work situations are not covered by these regulations because other regulations apply to them. However, these items need to be compatible with any other PPE provided.

ACTIVITY

What danger could arise from wearing a safety helmet back to front?

Personal protective equipment (PPE)

All equipment, including clothing for weather protection, worn or held by a person at work, which protects that person from risks to health and safety.

PPE should be used only when all other measures are inadequate to control exposure. It protects only the wearer while it is being worn and, if it fails, PPE offers no protection at all. The provision of PPE is only one part of the protection package; training, selection of the correct equipment in all work situations, good supervision, monitoring and supervision of its use, all play a part in the success of PPE as a control measure.

PROCEDURES FOR RECORDING UNSAFE SITUATIONS

If, during the course of any work, an electrician discovers an unsafe situation which is an immediate threat to life, they must be report it, in writing, to the responsible person or client immediately. If it is reasonable to make the situation safe, they should do so. For example, if an electrician discovers a plastic light switch that is damaged and live parts can easily be touched, it is not unreasonable for the electrician to replace that switch whether the client agrees to pay for the repair/replacement or not. Ultimately, it is the responsibility of the client or **responsible person** to decide if any repairs or replacements should be carried out. It is the duty of a **competent person** to advise them of the severity of any fault and what the statutory duty of the responsible person, or client, is and this must always be done in writing. More detail on this is covered in Learning outcome 2.

Responsible person

The person who is designated as the responsible person has delegated responsibility for certain aspects of a company's operational functions, such as fire safety, electrical operational safety or the day-to-day responsibility for controlling any identified risk such as *Legionella* bacteria.

Competent person

A person with relevant knowledge and experience of the work being undertaken, such as an electrician.

A fault is defined in BS 7671: 2008 Requirements for Electrical Installations (the IET Wiring Regulations) as: 'A circuit condition in which current flows through an abnormal or unintended path. This may result from an insulation failure or a bridging of insulation.'

PROCEDURES FOR RECORDING INFORMATION ON ELECTRICAL FAULT DIAGNOSIS

Electrical faults do not occur at convenient times, and fault diagnosis and repair will often be undertaken in difficult circumstances. Whatever the circumstances, the information gathered during the fault diagnosis and after repair must be recorded and retained for future use.

The Electricity at Work Regulations 1989 (EAW Regulations) is the principal legislation relating to work on electrical systems. In particular, Regulation 4(3) requires that: 'Work on or near to an electrical system shall be carried out in such a manner as not to give rise, so far as is reasonably practicable, to danger.'

Regulation 14 places a strict prohibition on working on or near live conductors unless:

a)   it is unreasonable for the equipment to be dead

b)   it is reasonable for the work to take place on or near the live conductor and

c)   suitable precautions have been taken to prevent injury.

In addition, employers are required under Regulation 3 of the Management of Health and Safety at Work Regulations 1999 to assess the risks to the health and safety of their employees while they are at work, in order to identify and implement the necessary precautions to ensure safety.

BS 7671, the Requirements for Electrical Installations, requires that, following the inspection and testing of all new installations, alterations and additions to existing installations or periodic inspections, an Electrical Installation Certificate (EIC), together with a Schedule of Test Results, should be given to the person ordering the work; this person is normally the client, **duty holder** or **responsible person**. Model forms for certification and reporting are contained in Appendix 6 of BS 7671.

Assessment criteria

2.1. State the procedures for reporting and recording information on electrical fault diagnosis and correction work

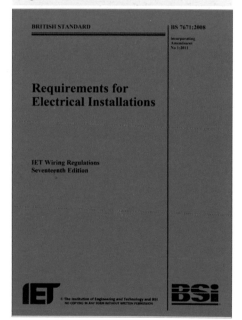

Duty holder

The person in control of the danger is the duty holder. This person must be competent by formal training and experience, and with sufficient knowledge to avoid danger. The level of competence will differ for different types of work.

Responsible person

The person who is designated as the responsible person has delegated responsibility for certain aspects of a company's operational functions, such as fire safety, electrical operational safety or the day-to-day responsibility for controlling any identified risk such as *Legionella* bacteria.

Using the correct forms

Faults would be recorded on an Electrical Installation Condition Report.

Electrical Installation Condition Report

The EICR should only be used for reporting on the condition of an electrical installation. It is intended primarily for the person who is ordering the work to be undertaken (the client, duty holder or responsible person for the installation) and anyone subsequently involved in additional or remedial work or additional inspections, to confirm, so far as is reasonably practicable, whether or not the electrical installation is in a satisfactory condition for continued service.

Each observation of a problem or concern relating to the safety of an installation should be given an appropriate classification code selected from standard classification codes as follows.

- *C1: danger present*. Risk of injury. Immediate remedial action required.
- *C2: potentially dangerous*. Urgent remedial action required.
- *C3: improvement recommended*.

An Electrical Installation Condition Report will have attached to it a set of guidance notes explaining the purpose of the report.

When an electrical fault has been diagnosed, identified and then repaired, the IET Wiring Regulations require that the circuit, system or individual piece of equipment be inspected and tested and that functional tests should be carried out. These inspections and tests must be carried out in accordance with Part 6 of the Regulations and the results recorded in accordance with Part 6, Chapter 63. The requirements that apply to fault diagnosis and repair are:

- 631.2 – completion of an EICR
- 631.4 – the EICR shall be compiled and signed by a competent person or persons
- 631.5 – the report can be produced in any durable medium, such as written hard copy, or by electronic means.

Depending on the extent of the fault and subsequent repair, either an EIC or a MEIWC will be issued to the person requesting the work. See pages 211–214 for more information on EIC or MEIWC.

USING FAULT CODES

Electrical installations degrade with time due to physical damage to switches, sockets and other fittings, together with deterioration of cables. The severity of degradation is more pronounced in installations within buildings where construction work is in progress, where adverse elements (such as corrosive chemicals or extreme temperatures) are involved or where maintenance has been poor. To ensure the safety of everyone who may come into contact with an electrical installation, such as users or people who undertake maintenance, it is vital that the installation is regularly inspected and tested to identify any faults or potential failures. Regulation 4(2) of the EAWR (1989) states: 'As may be necessary to prevent danger, all systems shall be maintained so as to prevent, so far as reasonably practicable, such danger.'

Regulation 4(2) recognises that the integrity of any electrical system can only be preserved in its 'initially as installed' condition if it is regularly maintained and repaired as necessary. This is no more or less than what one would expect with the use of any item of electrical or mechanical plant.

BS 7671 is the national standard in the United Kingdom for low-voltage electrical installations. It is non-statutory, but if wiring installation and maintenance work is undertaken in accordance with BS 7671 it is almost certain to meet the requirements of EAW Regulation 4(2). BS 7671 is largely based on the European Committee for Electrotechnical Standardisation (CENELEC) harmonised documents and is very similar to the current wiring regulations of other European countries. The Regulations deal with the design, selection, erection, inspection and testing of electrical installations operating at voltages up to 1000 V a.c.

As BS 7671 is non-statutory there is no legislative requirement which stipulates that a duty holder, responsible person or **person in control of the premises** must have or retain any kind of electrical inspection report. However, from a liability and safety perspective, it is advisable and provides proof that the requirements of the EAWR are being met. Rented properties and certain types of public place, such as theatres, restaurants, cinemas, clubs and hotels, are generally required to have some kind of report for insurance purposes and an EICR (as recommended in BS 7671) fulfils this role. In the event of a death or serious injury, resulting from failure to maintain the electrical integrity of an installation, some form of legal action could result.

Person in control of the premises

The person in control of the premises is likely to have physical possession of the premises or is the person who has responsibility for, and control over, the condition of the premises, the activities conducted on those premises and the people allowed to enter.

BS 7671 EICR

When an existing installation has been subjected to a periodic inspection and test, BS 7671 clause 631.2 recommends the provision of an EICR. The report should include details of the extent of the installation, and the limitations of the inspection and testing, together with records of inspection, the results of testing and a recommendation for the interval until the next periodic inspection. The EICR must be compiled and signed by a competent person or persons

Competent person

A person who possess sufficient knowledge, relevant practical skills and experience for the nature of the electrical work undertaken and is able at all times to prevent danger and, where appropriate, injury to him/herself and others.

ACTIVITY

Name the two other items of paperwork that should accompany an EICR.

and, in the context of low-voltage (LV) systems, BS 7671 contains definitions for '**competent person**', 'instructed person', 'skilled person' and 'ordinary person'.

As its title suggests, the EICR is a report, not a certificate. It relates to an assessment of the in-service condition of an electrical installation against the requirements of the issue of BS 7671 current at the time of the inspection, irrespective of the age of the installation.

The results, measurements and values taken during the inspection and testing are clearly recorded in a report. Appropriate recommendations, if applicable, are made to rectify any damage, deterioration or defects, dangerous conditions and non-compliance with the requirements of the Regulations that may give rise to danger, along with any limitations to the inspection and testing, and the reasons for these.

The EICR contains 11 sections, which are identified alphabetically from A to K. Section K, 'Observations' has three columns to be completed. Observation(s) are entered in the first column. The second column requires a classification code (C1, C2 or C3) with reference to the observation(s) and the third column is used to record whether or not further investigation is required.

Code C1 – danger present

This code indicates that there is a risk of injury and that immediate remedial action is required to remove the dangerous condition.

Code C1 allows those carrying out an inspection to report to the client or responsible person that a risk of injury exists, which could be, for example, accessible live conductors due to damage, poorly modified enclosures or removed maintenance panels. Incorrect polarity would also attract a Code C1 as it may allow conductive parts, not normally expected to be live, to become live.

A reported Code C1 warrants immediate action to be taken. This involves immediately informing the client or responsible person for the installation, both verbally and in writing, that a risk of injury exists. A detailed explanation of this risk should be recorded on the report, together with details of any verbal and written warnings. If possible, dangerous situations should be made safe or rectified before further work or inspections are carried out.

Code C2 – potentially dangerous

This code indicates that urgent remedial action is required. The EICR should declare the nature of the problem, but not the remedial actions required.

The phrase 'potentially dangerous' indicates a risk of injury from contact with live parts following a 'sequence of events'. A sequence of events could mean that an individual would need to move, open or gain access to live parts when undertaking a daily task that would not normally be expected to give access to live parts.

An example of this would be an isolator with a damaged casing, in a locked cupboard. This might leave exposed live parts. But these could only be accessed with the use of access equipment such as a specialist tool or key. An individual would need to gain access to the cupboard before coming into contact with live parts, but nevertheless the potential for risk of injury is high.

The lack of an adequate earthing arrangement for an installation, the use of utility pipes as the means of earthing or an undersized earthing conductor (in accordance with BS 7671 Regulation 543.1.3) would also warrant a Code C2 because a primary fault would be needed in order for these scenarios to become potentially dangerous.

Code C3 – improvement recommended

This code implies that, while the installation may not comply with the current edition of the Regulations, it does comply with a previous edition and is deemed safe, although improvements could be made.

An example could be where an installation not modified in recent years has cables concealed in a wall that are not provided with additional protection by an RCD.

ACTIVITY

An installation is 30 years old and, apart from minor wear and tear, is satisfactory. It does, however, have rewireable fuses in the consumer unit. Outline what changes you may suggest following replacement of old pendants.

Which codes, if any, should you use in this situation?

ASSESSMENT GUIDANCE

Remember that just because an installation does not comply with current regulations this does not make it unsafe. Most installations will still comply with the regulations in force when they were installed.

ACTIONS TO BE TAKEN IN RESPONSE TO DANGER CODES

An EICR is intended primarily for the person who is ordering the work to be undertaken (the client, duty holder or responsible person for the installation) and anyone subsequently involved in additional or remedial work or additional inspections to confirm, so far as reasonably practicable, whether or not the electrical installation is in a satisfactory condition for continued service.

Code C1

Wherever practicable, items classified as 'danger present' (C1) should be made safe on discovery. The duty holder or responsible person ordering the report should be advised of the action taken and on the necessary remedial work to be undertaken. If that is not practical you must take other appropriate action, such as switching off and isolating the affected parts of the installation to prevent danger.

Secure

This can best be achieved by locking off with a safety lock – such as a lock with a unique key. The posting of a warning notice also serves to alert others to the isolation.

There are two separate and distinct requirements for making a section of an electrical system safe when a Code C1 has been observed and recorded.

- Cutting/switching off the supply – depending on the equipment and the circumstances, this may be no more than carrying out normal functional switching (on/off) or emergency switching by means of a stop button or a trip switch.

- Isolation – this means the disconnection and separation of the electrical equipment from '*every source of electrical energy* in such a way that this disconnection and separation is **secure**'.

Secure isolation using suitable locks and labelling

As has already been stated, the purpose of a condition report is to confirm, so far as is reasonably practicable, whether or not the electrical installation is in a satisfactory condition for continued service.

Code C2

For items classified as 'potentially dangerous' (C2), the safety of those using the installation may be at risk and it is recommended that a competent person undertakes any necessary remedial work as a matter of urgency.

If this potentially dangerous situation can be rectified by an addition or a simple alteration to the installation, the client will probably request that this is carried out. It should be remembered that this is not part of the initial agreement. If the work does not involve major system changes, such as the installation of a new circuit, then a MEIWC should be issued. Remember an EIC is to be used only for the initial certification of a new installation or for an alteration or addition to an existing installation where new circuits have been introduced.

Code C3

For items classified as Code C3, 'improvement recommended', the inspector has found some problems that are not immediate risks but may become risks if not improved. One example may be that the inspection has revealed an apparent deficiency that could not, due to the extent or limitations of the inspection, be fully identified.

A further examination of the installation will be necessary, to determine the nature and extent of the apparent deficiency. Such investigations should be carried out as soon as possible.

For safety reasons, the electrical installation will need to be reinspected at appropriate intervals by a competent person. The recommended date by which the next inspection is due should be included in the report, under 'Recommendations', and on a label near to the consumer unit or distribution board.

On completion of the inspection, the client, duty holder or responsible person ordering the report should receive the original report and the inspector should retain a duplicate copy. The original report should be retained in a safe place and be made available to any person inspecting or undertaking work on the electrical installation in the future. If the property is vacated, the report will provide the new owner or occupier with details of the condition of the electrical installation at the time the report was issued.

THE PROCESS OF FAULT RECTIFICATION

Electrical faults do not occur at convenient times, and fault diagnosis and repair is often undertaken in difficult circumstances. For the duty holder or person responsible for the electrical installation, it can be difficult to determine the most suitable arrangements for fault rectification, due to the potential for danger to people and equipment, time delay and lost production.

The consequences of a fault on an electrical installation can be considerable: lost production due to the unavailability of a machine, lost business where goods, products or services are not being made available to a customer and data loss if, for example, the backup UPS fails to function. Therefore, a suitable process needs to be established in order to keep inconvenience and costs to an acceptable level, without compromising the safety of users of the system and electricians undertaking the fault repair.

Having identified the fault, the most suitable rectification process should be identified. This could be as simple as correcting a loose connection at an accessory terminal or undertaking a repair by replacing a faulty conductor.

Assessment criteria

2.2 State the procedures for informing relevant persons about information on electrical fault diagnosis and correction work and the completion of relevant documentation

 SmartScreen Unit 308
Handout 22

Agreeing the process of fault rectification

Often the solution will be fairly straightforward but, in some circumstances, options must be discussed and agreed with the duty holder or responsible person. Some or all of the following issues may be open to consideration.

- What is the cost of replacement?
- What is the availability of suitably trained staff?
- Is it possible to replace a lesser number of components?
- What is the cost comparison of replacing or repairing a component?
- What is the potential disruption to the manufacturing process or computer availability?
- What is the availability of replacement parts, if required?
- Can other parts of the system be energised whilst the repair is carried out?
- What is availability of emergency or standby supplies?
- What is the anticipated 'down time' while a repair is carried out?
- Is there a requirement for a continuous supply?
- Will out-of-hours working be required?
- Are there any legal or personnel issues to be considered (warranties, contracts)?

Once the decision on how the fault rectification is to proceed has been made and agreed with the duty holder or responsible person, a work activity plan should be prepared. This, once again, must be agreed with the duty holder. The plan should include a safe system of work and requirements for circuit outages.

Unless the fault diagnosis and rectification are being undertaken by an in-house electrician, site access must be granted by the duty holder or responsible person. This could involve a site induction with site access passes being issued. There might also be a logging-in procedure (visitors' book) required for each visit.

The preceding arrangements are all concerned with a fault or supply failure on the client's system. However, the initial fault diagnosis may determine that the problem is connected with the system of the distribution network operator (DNO). If this is the case, the responsibility for the fault rectification rests with the network operator. Under these circumstances, it is good business practice to advise the client on how to deal with the DNO, requesting the provision of temporary standby supplies in the form of a backup generator if required.

The work activity should be managed with appropriate technical support and resource allocations to ensure that an effective repair is completed within the agreed timescales.

ASSESSMENT GUIDANCE

An effective maintenance system will often prevent a number of faults before they can become major problems.

PLANNING AND AGREEING PROCEDURES

Before fault diagnosis is carried out, the safe-working arrangements must be discussed and agreed with the client and/or the duty holder (or responsible person for the installation) in a clear, concise and courteous manner. The testing and inspection procedure must be a planned activity as it will almost certainly affect people who work or live in the premises where the installation is being tested.

This ensures that everyone who is concerned with the work understands what actions need to be taken, such as:

- which areas of the installation may be subject to disconnection
- anticipated disruption times
- who might be affected by the work
- health and safety requirements for the site
- which area will have restricted access
- whether temporary supplies will be required whilst the fault diagnosis is underway
- the agreement reached on who has authority for the diagnosis and repair.

It may be that a specific person has responsibility for the safe isolation of a particular section of an installation and that person should be identified and the isolation arrangements agreed. By entering into dialogue with the client before work commences, the potential for unforeseen events will be minimised and good customer relations will be fostered.

For example, in an office block where the electrical installation is complex and provides supplies to many different tenants located on a number of floors, the safe isolation of a sub-circuit for testing purposes may require a larger portion of the installation to be turned off initially. In order to achieve this with minimal disruption, an agreement must be reached between the competent person tasked to carry out the work and the person responsible for the installations affected. This responsible person could be the office manager, the designated electrical engineer for the site or, in some cases, the landlord of the building.

Everyone involved in the work (for example, client, electrician and those in the workplace) has a responsibility for their own health and safety and that of others who may be affected by the work; communication between all parties will ensure compliance with the respective health and safety requirements.

KEY POINT

You should appreciate the difference between the duty holder and the responsible person.

The person in control of the danger is the *duty holder*. This person must be competent by formal training and experience and with sufficient knowledge to avoid danger. The level of competence will differ for different items of work.

The person who is designated the *responsible person* has delegated responsibility for certain aspects of a company's operational functions such as fire safety, electrical operational safety or the day-to-day responsibility for controlling any identified risk such as *Legionella* bacteria.

ACTIVITY

Using any source of reference material, obtain a copy of a permit to work and determine when a PTW might be used on a low voltage installation.

Understand how to complete the preparatory work prior to fault diagnosis and correction work

Assessment criteria

3.1 Specify safe working procedures that should be adopted for completion of fault diagnosis and correction work

SAFE WORKING PROCEDURES FOR FAULT RECTIFICATION

The possibility of touching live parts is increased during fault finding and repair, when conductors at dangerous voltages are often exposed. This risk can be reduced if work is done while the equipment is isolated from any dangerous source of electrical supply. However, this may not always be possible and, if this is the case, procedures should be followed to prevent contact with any hazardous internally produced voltages.

Before fault finding and repair work commences, anyone carrying out the work must be aware of the following relevant legal requirements:

- the Health and Safety at Work etc Act 1974 in relation to safe systems of work

- the Electricity at Work Regulations 1989 regarding work on or near to an electrical system

- the Management of Health and Safety at Work Regulations 1999 which covers risk assessment.

The method statement

Once the risk assessment in accordance with the Management of Health and Safety at Work Regulations has been carried out, a safe system of work, sometimes called a method statement, can be prepared to enable the work to be undertaken in a safe manner.

The safe system of work or method statement will include such items as:

- who is authorised to undertake testing and, where appropriate, how to access a test area and who should not enter the test area

- arrangements for isolating equipment and how the isolation is secured

- provision and use of personal protective equipment (PPE) where necessary

- the correct use of tools and equipment

- the correct use of additional protection measures, for example, flexible insulation that may need to be applied to the equipment under test while its covers are removed

- use of barriers and positioning of notices

- safe and correct use of measuring instruments

- safe working arrangements to be agreed with client, duty holder or responsible person

- how defects are to be reported and recorded

- instructions regarding action to be taken if unsafe situations are identified.

Guidance on live and dead working can be found in the HSE publication *Memorandum of guidance on the Electricity at Work Regulations 1989* (HSR 25) and can be downloaded free of charge from the Health and Safety Executive website: www.hse.gov.uk.

Safe working procedures should be reviewed regularly, to make sure that they are being followed and are still appropriate for the work that is being carried out. If any changes are made to the procedures, all people who are involved in the fault diagnosis regime should be given relevant instruction and training.

A permit to work (PTW) procedure, as described on page 151 of Learning Outcome 1, may be required if the installation is complex in nature.

Planning and agreeing procedures

Before fault rectification is carried out, the safe working arrangements must be discussed and agreed with the client and/or the duty holder (or responsible person for the installation) in a clear, concise and courteous manner. The procedure must be a planned activity as it will almost certainly affect people who work or live in the premises where the installation is being tested. This ensures that everyone who is concerned with the work understands what actions need to be taken, such as:

- which areas of the installation may be subject to disconnection

- anticipated disruption times

- who might be affected by the work

- health and safety requirements for the site

- which area will have restricted access

- whether temporary supplies will be required whilst the fault repair is underway

- the agreement reached on who has authority for the repair.

ASSESSMENT GUIDANCE

Never assume a circuit is dead. Carry an approved voltage indicator with you at all times to test the circuit.

ACTIVITY

A lathe is driven by a three-phase motor. It is one of a number of machines in a factory making automobile components. The motor is believed to have developed an internal fault. Complete a method statement for the fault-finding operation on the machine.

It may be that a specific person has responsibility for the safe isolation of a particular section of an installation, in which case that person should be identified and the isolation arrangements agreed. Entering into dialogue with the client, before work commences, will minimise the potential for unforeseen events and foster good customer relations.

For example, in an office block where the electrical installation is complex and provides supplies to many different tenants located on a number of floors, the safe isolation of a sub-circuit for testing purposes may require a larger portion of the installation to be turned off initially. To achieve this with minimal disruption, an agreement must be reached between the competent person tasked to carry out the work and the person responsible for the installations affected. This responsible person may be the office manager, the designated electrical engineer for the site or, in some cases, the landlord of the building.

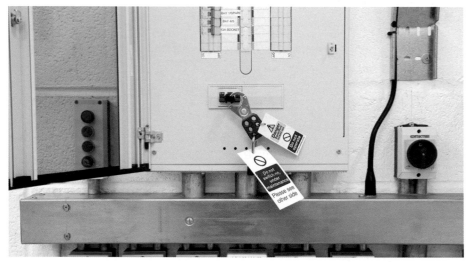

Isolation of an installation should be part of a planned and agreed procedure

Everyone involved in the work (for example, client, electrician and those in the workplace) has a responsibility for their own health and safety and that of others who may be affected by the work. Communication between all parties will ensure compliance with the respective health and safety requirements.

A LOGICAL APPROACH FOR LOCATING ELECTRICAL FAULTS

Locating faults which occur on an electrical system requires a logical approach. This approach includes:

- identifying the symptoms
- collecting and analysing data
- use of information such as drawings, certificates, instructions
- checking maintenance records
- using experience
- inspecting, checking and testing.

In order to follow the logical approach as listed above, the approved electrician must have:

- an approved safe working procedure or method statement as detailed on page 148, including a risk assessment

- a thorough knowledge and understanding of the electrical installation and electrical equipment (including access to site-specific circuit diagrams and drawings, manufacturers' information and operating instructions, design data, copies of previous Schedules of Inspection or certificates, installation specification and maintenance records)

- information relating to the fault, including any available knowledge of events leading up to the fault (such as unusual smells or observations); this can be obtained from users, the duty holder or the responsible person

- details of personnel relevant to the operation of the installation.

If a fault has occurred, it is likely that users of the system will have reported a loss of supply or changes in the way the system operates; they will probably be able to identify the areas affected. Determining which protective device has operated (main circuit breaker, individual circuit breaker or residual current device) will narrow the search area and a flow diagram can often assist in this process.

Questions to ask users include the following:

- Which parts of the system are without supply – the whole installation or a particular item of equipment?

- Is the failure a permanent problem or does it only happen at certain times or on certain days?

- What are the events that led up to the fault occurring?

Checks that an approved electrician can undertake

Visual inspection

A visual inspection is a quick way of identifying defects, damage or deterioration of the electrical installation. Although this check is commonly called a visual inspection, other senses such as smell, touch and hearing are also used. Touch can be used to test conductors to see if they are loose at joints and connections and smell can be used to identify overheating of thermoplastic material caused by a high resistance fault. Hearing can be used to detect arcing at a loose termination.

Here is an initial checklist for fault finding.

- Joints and connections – are conductors loose or is a connection missing?
- Conductors – do they show signs of overheating?
- Flexible cables and cords – are there signs of exposed insulation or conductors?
- Switching devices – any signs of overheating?
- Protective devices – have they operated?
- Enclosures and mechanical protection – any signs of overheating or physical damage?

Check the supply voltage

Ask these questions.

- Is the complete system, together with all circuits, dead?
- Is a particular section of the installation dead?
- Has an individual piece of equipment failed?

Use an approved voltage tester to check for a supply voltage. If there is no incoming supply from the distribution network operator (DNO), this may be due to a main fuse failure, circuit breaker operation or a fault on the DNO network. It will be necessary to contact the distribution company to effect a repair.

Checking the supply voltage

Check the protective devices

Ask these questions.

- Has the main circuit breaker tripped?
- Is the system overloaded; have unauthorised items of equipment been connected to the system?
- Is the wiring in the distribution board damaged in some way, or is there a short circuit somewhere in the system between live and neutral conductors?

SmartScreen Unit 308
Handout 15 and Worksheet 15

ASSESSMENT GUIDANCE

When you dismantle equipment make sure you label, if necessary, all parts and store them securely.

Check individual items of equipment

Check whether the fault is restricted to one item of equipment or to the radial circuit feeding a particular piece of equipment.

For example, if the protective device operates every time a large item of equipment is switched on, it is probably the equipment that is faulty rather than the supply cable. A cable fault would normally cause the protective device to operate even when the equipment is switched off. Carry out a visual inspection of the equipment, looking for signs of damage or burning. If no physical signs of damage are apparent, undertake an insulation resistance test on the item of equipment.

Remember, before testing and fault diagnosis is carried out, the safe working arrangements must be discussed and agreed with the client, duty holder or responsible person for the installation. Everyone concerned should understand what the work entails, which areas of the installation may be subject to disconnection and who might be affected by the testing.

CAUSES OF ELECTRICAL FAULTS

A fault is defined in BS 7671: 2008 Requirements for Electrical Installations as: 'a circuit condition in which current flows through an abnormal or unintended path. This may result from an insulation failure or a bridging of insulation'. A fault is not a normal occurrence and the abnormal current flow could be caused by one of the following:

- poor system design
- designer's specification incorrectly implemented
- installation workmanship to a poor standard
- poor maintenance or neglect of equipment or installation
- misuse, abuse or deliberate ill-treatment of equipment.

Poor system design

The proper specification and design of electrical systems and equipment is fundamental to the safe operation of installations. The designer of a low-voltage system will need a good working knowledge of the current edition of BS 7671, associated IET Guidance Notes and codes of practice, technical and equipment standards. They should also understand the duties imposed by Regulations 4 to 15 of the EAWR and the technical reasons behind them.

Failure to interpret the design correctly

However good the design of a system, it has to be installed correctly to function as intended. Those concerned with installing the system must be competent to ensure that the designer's specification is correctly implemented and all workmanship is to a high standard.

ASSESSMENT GUIDANCE

It could be that the original installation is perfectly acceptable but changes to the building layout may adversely affect the installation.

ACTIVITY

What is a definition of *competence*?

Installation workmanship to a poor standard

Typical areas of poor installation practice are:

- poor termination of conductors, which creates overheating due to poor electrical contact
- loose bushings and couplings, which may lead to poor earth continuity and risk of shock
- use of incorrectly sized conductors, leading to overheating
- cable damage that occurs when drawing in cables
- heavily populated trunking or conduit, leading to overheating
- incorrect connections at components, which may lead to crossed polarity (so that a circuit remains live although the protective device has operated).

Those concerned with installation work need to know not only *how* to do a job, but *why* it is important to do it in the specified way.

Misuse of equipment

EAW Regulation 4(3) requires that: 'every work activity, including operation, use and maintenance of a system and work near a system, shall be carried out in such a manner as not to give rise, so far as is reasonably practicable, to danger and users (the duty holder) have a responsibility to use equipment in the manner for which it was intended.' Deliberate or unintentional misuse, such as overloading a radial circuit supplying socket outlets with too many electric heaters, could be seen as misuse.

Abuse or deliberate ill-treatment of equipment

Abuse or deliberate ill-treatment of equipment, such as removal of barriers or covers, may result in a short circuit because of contact between live conductors.

COMMON FAULTS

An electrical fault can reveal itself in a number of ways or in a combination of those ways, for example:

- complete loss of power throughout the whole installation
- partial or localised loss of power
- failure of an individual component.

Failure of an individual component

Faults in components such as a switches, contactors, time switches and photo electric cells may result in particular circuits not operating as was intended. For example, heaters that are programmed to operate via a time switch can only be turned off by operation of the protective device.

Total failure of a piece of equipment

An example might be the failure of a heating element in a fire or shower, resulting in an earth fault or short circuit.

Insulation failure

An insulation fault can allow current leakage. As insulation ages, its insulating properties and performance deteriorates; this is accelerated in harsh environments, such as those with temperature extremes and/ or chemical contamination. This deterioration can result in dangerous conditions for people and poor reliability in installations. It is important to recognise this deterioration so that corrective steps can be taken.

Excess current

If excess current is drawn by a circuit, protective devices will operate frequently. This could be caused by too many items of equipment being connected at the same time.

Short circuits

These occur when two or more components, which would normally be separated by insulation or barriers, come in contact with each other. For example, in an overload condition the circuit conductors are carrying more current than the manufacturer's design specification allows and, if this is allowed to continue indefinitely, the conductors will become very hot, causing deterioration of the electrical insulation surrounding the live conductors. This may eventually lead to breakdown of the insulation and a short circuit. In this overload situation, it is likely that the protective device, if specified correctly, will operate before a short-circuit condition is reached. Continual operation of a protective device should be investigated immediately, by isolating the circuit or carrying out an insulation resistance test, or by using a clamp meter or clip-on meter to check load readings.

Open circuits

An open circuit generally occurs when there is a break in the circuit. This could be the result of a broken conductor, a loose connection caused by excessive mechanical stress or a component that has failed due to age, damage, overloading, suitability for continued use and overheating.

Transient overvoltage

This is caused by heavy current switching and distribution network operator (DNO) system faults. Occasionally, large internal loads may cause such transient overvoltages.

The designer of a system must ensure that all system components are selected according to their intended use and are appropriately rated. Two main factors must be considered:

■ switchgear, fuse gear and cables must be of adequate rating to carry both normal and any likely fault current safely

■ the equipment should meet appropriate British, European or International Standards or CENELEC Harmonisation Documents (HDs) in order to withstand prospective fault conditions in the system.

However, design work does not cater for abnormal transient voltages caused by distribution network operator (DNO) faults on the supply network, heavy current switching that causes voltage drops, lightning strikes on overhead line conductors, electronic equipment and earth fault voltages. In some installations, such as IT networks, where transient voltages can cause problems, additional measures are included. Filters added to IT networks can suppress transient voltages and provide stabilised voltage levels.

High-resistance faults

These faults usually occur in a circuit where a cable or conductor is poorly joined in an accessory such as a socket outlet, switch, light fitting, or junction box. The most common cause is simply a loose screw terminal connection, because the connection was not made sufficiently well in the first place, debris is present in the connection (for example pieces of stripped insulation), or it might be that the connection has become loose over time. Connections can work loose through the normal thermal cycling of a circuit. If a circuit routinely carries a significant proportion of its maximum design current, the cable will be subject to repeated heating and cooling cycles. This can result in expansion and contraction, which can loosen the terminal screws. Vibration can also affect joints and connections, and that is why connections for generators need to be crimped or brazed.

WHERE ARE ELECTRICAL FAULTS FOUND?

Electrical faults can occur in a range of different areas and components. The following section is not exhaustive but highlights the major areas where faults may occur.

Wiring systems

Wiring faults do occur in low-voltage systems but cable damage is not a common occurrence if the installation has been constructed in accordance with Chapter 52 of BS 7671.

However, if the installation has not been constructed to a high standard, damage may occur and a latent fault may exist, causing a failure at a later date. PVC insulated and sheathed wiring is extensively

Assessment criteria

3.5 Specify the types of faults and their likely locations in:

■ wiring systems
■ terminations and connections
■ equipment/accessories (switches, luminaires, switchgear and control equipment)
■ instrumentation/metering

used for wiring in domestic and light industrial installations, and the very act of installing cable runs can result in mechanical damage to the cable due to impact, abrasion, compression or penetration (which must be minimised at all times). Where cables have to be turned or bent, the radius should be such that it does not damage the conductors. Where cables are installed in walls behind plaster or in a stud wall, mechanical protection is only required if the depth of the cable is less than 50 mm, or if they are not placed within accepted zones that are 150 mm wide around wall edges and ceiling–wall joints, or if the cables are not run horizontally or vertically to an accessory point. If it is not possible to provide such protection and the installation is used by people with no technical background, additional protection must be provided by the installation of a residual current device.

When designers calculate the conductor sizes for installations, based on load factors, they also specify how the cables should be installed. Any deviation from the specification could have an adverse effect – cable current-carrying capacities will be reduced by such factors as grouping, surrounding the cables with thermal insulation and siting the cables in a hot environment that will cause overheating. Bear in mind that this could take place after the installation was originally commissioned. Poor installation practice could also result in cables being subjected to compression, pinching or abrasive damage.

Surface cabling in industrial workplaces, or where there is movement of materials or goods, must be protected from mechanical damage. This may be achieved by the use of cables incorporating mechanical protection (such as steel wire armoured or mineral insulated cable) or by the installation of cables in steel conduit or trunking, or by locating the cables where they will not be liable to damage.

Wiring faults often occur where cables are jointed or terminated. A termination is a connection between a conductor and other equipment (for example distribution boards or fixed appliances), or an accessory. Connections between conductors are usually termed 'joints'.

A poorly constructed joint or termination may give rise to arcing with the passage of load current. This will result in localised heating (and an increased fire risk) and progressive deterioration of the joint or termination. There will be evidence of pitting, arcing, carbon build-up or corrosion that could lead to eventual failure and an open circuit. Loose connections developing into a short circuit are often accompanied by audible sounds such as 'crackling' or 'fizzing' and by referred problems in adjacent parts of the installation.

When a fault occurs elsewhere on an associated part of an electrical system, a poor joint or termination may constitute a hazard by limiting the flow of fault current, thereby causing delayed operating of the circuit protective device. In an extreme case, the passage of high fault current may cause the defective joint to burn into an open-circuit fault before the circuit protective device has time to operate, with the result that the original fault remains uncleared. This is particularly hazardous in the case of an earth fault, since exposed conductive parts may become and remain live at the system voltage, giving an increased risk of electric shock.

Equipment and accessories

Faults can also occur in equipment and accessories such as switches, sockets, control equipment, motor contactors and electric appliances or at the point of connection with electronic equipment. Contacts that make and break a circuit are a potential source of failure, depending on the amount of operation they are subjected to. Socket outlets and isolating switches that control fans in kitchens or bathrooms, or showers in bathrooms, that are subjected to regular use with capacity loads may fail due to overheating and loose connections.

Failure of an individual component, such as an electric shower or immersion heater element, may lead to an open-circuit fault. Photo-electric cells and time switches may fail in the 'closed position' and then do not operate as intended. Faulty RCDs may cause nuisance tripping. The correct operation of a suspected faulty RCD would need to be determined by the use of an approved RCD tester (Unit 307, page 103).

Modern RCD testers include a facility called a ramp test. The ramp test slowly increases the tripping current until the RCD trips. The display is in mA and will identify if an RCD is susceptible to nuisance tripping or is out of specification. It can also be used to indicate which circuits have high leakage currents and can be used when there is evidence of frequent trips on a particular RCD.

Under certain circumstances these tests can result in potentially dangerous voltages appearing on exposed and extraneous conductive parts within the installation. Therefore, suitable precautions must be taken to prevent contact by persons or livestock with any such part. Other users of the building should be made aware of the tests being carried out and warning notices should be posted as necessary.

Cables that are terminated at an accessory contained in, or passing through, a metal box should be bushed with a rubber grommet to prevent abrasion of the cable.

Instrument and metering panels

Faults can occur in instrument and metering panels as a result of a faulty component, such as a burnt-out current transformer (CT) or voltage transformer (VT), or as a result of faulty test probes being applied to the instrument.

ACTIVITY

Remember that all equipment has some earth leakage, however small. It may be the sum of many leakages that causes a main RCD to trip which is controlling several circuits. What would you do to resolve this problem?

Protective devices

These may be subject to loose connections. If an incorrect device was selected for the circuit and does not offer the correct protection or discrimination, a fault may result.

Luminaires

Light fittings often fault because the lamp has expired. Fluorescent fittings (discharge lighting) grow dimmer with age, and may even begin to flicker or to flash on and off. These are warning signals and the necessary repairs should be made as soon as there is any change in the lamp's normal performance. A dim tube usually requires replacement, and failure to replace it can strain other parts of the fixture. Likewise, repeated flickering or flashing will wear out the starter, causing the insulation at the starter to deteriorate.

Flexible cords

Many faults occur when flexible cords are terminated because they are:

- of too small cross sectional area for the load

- not adequately anchored to reduce mechanical stresses

- not suitable for the ambient temperature at the point of termination.

It should be remembered that components are less likely than poor installation practices to cause a fault on an installation. Therefore, adherence to EAWR and BS 7671 is vital to ensuring that the electrical installation remains fault free.

HAZARDS OF FAULT DIAGNOSIS

The Electricity at Work Regulations 1989 (EAWR) require those in control of part or all of an electrical system to ensure that it is safe to use and that it is maintained in a safe condition. The process of identifying and rectifying faults will inevitably involve exposure to some hazards and these must be dealt with in accordance with the requirements of the EAWR.

Minimising hazards of lone working

Fault diagnosis, which may require some form of live electrical work, is often undertaken by people working alone and without close or direct supervision. There is no specific requirement contained in the

> **ASSESSMENT GUIDANCE**
>
> An old choke nearing the end of its life may become noisy as the laminations loosen.

> **Assessment criteria**
>
> **3.6** State the special precautions that should be taken with regard to the following:
> - lone working
> - hazardous areas
> - fibre-optic cabling
> - electro-static discharge (friction, induction, separation)
> - electronic devices (damage by over voltage)
> - IT equipment (eg shutdown, damage)
> - high frequency or capacitive circuits
> - presence of batteries (eg lead acid cells, connecting cells)

Electricity at Work Regulations for two people to be present in live-work situations. It is the responsibility of the duty holder to assess the need for a second person in terms of the nature of the danger and other precautionary measures to be adopted (such as role planning and relevant training). If a second person is *not* present, consideration must be given to special procedures for the lone worker such as:

- making sure that every site visit is recorded in a visitor log book
- keeping in contact with the employer by making regular telephone calls
- signing out when leaving the site for any reason
- making contact with the site owner on returning to site, to confirm that there have been no material changes to the system since the last visit.

It is likely that an electrician visiting a site for fault rectification purposes will be required to undergo some form of induction training which will explain the requirement to observe site-specific elements appropriate to their own work and/or site-specific activities, such as:

- site hazards and risks, for example, open excavations, presence of overhead power lines, confined spaces and hazardous areas
- fire risks and site fire procedures
- area of work that will require specific authorisation to proceed such as a permit to work
- restricted areas and the reasons for the control measures in place.

Minimising hazards from static electricity

Static electricity is produced by the build-up of electrons in weak electrical conductors or insulating materials. These materials may be gaseous, liquid or solid and may include flammable liquids, powders, plastic films and granules.

A simple example of static electricity is a person walking across a carpeted floor, who builds up a static charge because of the friction of their shoes on the carpet. When the person touches an object that is either uncharged or that has an opposite charge, a short pulse of very high-voltage static charge is released. This can cause damage to sensitive circuitry in computers, laptops and communication systems. Anti-static carpet sprays can be used to combat this problem.

Large static charges may develop on drive belts between motors and machinery. Examples include laminate manufacturing processes and paper making processes involving large rolls of toilet tissue or kitchen roll running at high speeds. If not diverted through proper earthing, the built-up static charges may discharge suddenly and cause damage to electronic equipment such as proximity switches and programmable logic controllers. Earthed anti-static strips, which look similar to tinsel, can be strategically positioned to continuously discharge these unwanted voltages.

In areas containing explosive materials, great care must be taken to prevent static discharges, since a spark could set off a violent explosion. The use of large-diameter pipes for the transfer of liquids reduces the flow rate and, hence, the build-up of static charge. Airborne fibres, dust or flecks of paper – such as those produced in industrial processes – should be removed at source and not be allowed to accumulate as this may create a fire risk if static is likely.

The static effect is increased in environments with low humidity and in buildings with air conditioning and high levels of heating, such as computer suites. This should always be considered when undertaking electrical work in such areas.

Minimising hazards from fibre-optic cabling

As more fibre-optic cable systems are introduced into installations, the maintenance and fault-finding demands increase. Probably the greatest hazard to deal with, for approved electricians, is fibre-optic cable's benign appearance. It doesn't carry electricity, so there is no electrocution risk; it isn't a source of heat or combustion so there is no fire risk; and it's not possible to know when it is operating.

Energy is transferred along fibre-optic cables as digital pulses of laser light and, therefore, direct eye contact with the light should be avoided. However, damage to the eye caused by looking into a damaged fibre-optic cable is rare as the broken surface of the cable tends to scatter the light coming through it.

A more serious hazard of optical-fibre work comes from the fibres themselves. Fibres are pieces of optical glass which, like any piece or sliver of glass, can cause injury. Hazards arise when cables are opened and fibres penetrate the hands or are transferred to the respiratory or alimentary system. Therefore, suitable PPE (such as gloves and face masks) should be used when working with fibre-optic cabling.

Fibre-optic cables have a similar appearance to steel wire armour cables (SWA) but are much lighter. They should be installed in the same way and given a similar level of protection as steel wire armour cables. Fibre-optic terminations should always be carried out in accordance with the manufacturer's instructions.

ACTIVITY

Static electricity can be generated when a person simply walks across a room. Clothing movement causes problems, especially when man-made fibres are involved. How could static problems be prevented in an operating theatre?

Electricians should use appropriate PPE when dealing with fibre-optic cables

KEY POINT

Fibre-optic cables are termed Band I circuits when used for data transmission and must be segregated from other mains cable. Band I covers installations where the voltage is limited for operational reasons (telecommunications, signalling, bell, control and alarm installations). Extra-low voltage (ELV) will normally fall within Band I.

Minimising hazards from electronic devices

The use of electronic circuits in all types of electrical equipment is now commonplace, with components being found in motor starting equipment, control circuits, emergency lighting, discharge lighting, intruder alarms, special-effect lighting and dimmer switches, and domestic appliances. Electrical installations usually operate at currents in excess of 1 A and up to many hundreds of amperes; all electronic components and circuits are low voltage and usually operate in mA or µA. This fact must therefore be considered when the choice of electrical test equipment is made.

The working voltage of a standard insulation resistance tester (250 V/500 V) may cause damage to any of the electronic devices described above, if they are not rated for this voltage. When carrying out insulation resistance tests as part of a prescribed test or after fault repair all electronic devices must be disconnected. Should resistance measurements be required on electronic circuits and components, a battery-operated ohmmeter with high impedance must be used.

Minimising hazards from high-frequency or capacitive circuits

Capacitors that are used in fluorescent fittings and single-phase motors are a potential source of electric shock arising from:

- the discharge of electrical energy retained by the capacitor unit(s) after they have been isolated

- inadequate precautions to guard against electric shock as a result of any charged conductors or associated fittings

- charged capacitors that are inadequately short circuited

- equipment retaining or regaining a charge.

Live working is to be avoided where possible, but any equipment containing a capacitor must be assumed to be live or charged until it is proven, beyond any doubt, that the circuits are dead and the capacitors are discharged and are unable to recharge.

ASSESSMENT GUIDANCE

Capacitors are not only used in equipment but as stand-alone units for power factor correction. These need to be isolated and discharged to be safe.

Minimising hazard from storage batteries

Batteries are used to store electrical energy. Their ability to provide instant output makes them a regular source of standby power for emergency lighting, emergency trip coils and uninterruptable power supplies (UPS) at computer, data storage and telecommunication facilities, for example.

They can prove invaluable for maintaining supplies where a supply interruption could cause business issues. There is, however, risk attached to these larger battery systems which, if used incorrectly, can be dangerous and can cause explosions.

There are two main classes of battery.

- *Lead–acid batteries* are the most common large-capacity rechargeable batteries. They are fitted extensively in UPSs of computer and communication facilities, and process and machinery control systems.

- *Alkaline rechargeable batteries*, such as nickel–cadmium, nickel–metal hydride and lithium ion, are widely used in small items such as laptop computers, but large-capacity versions of these cells are now being used in UPS applications.

Batteries can explode if used incorrectly

The two main hazards from batteries are described below.

- *Chemical* – lead–acid batteries are usually filled with electrolytes such as sulphuric acid or potassium hydroxide. These very corrosive chemicals can permanently damage the eyes and produce serious chemical burns to the skin. If acid gets into the eyes, they should be flushed immediately with water for 15 minutes, and prompt medical attention should be sought. If acid gets on the skin, the affected area should immediately be rinsed with large amounts of water and prompt medical attention should be sought if the chemical burn appears serious. Emergency wash stations should be located near lead–acid battery storage and charging areas. Under severe overcharge conditions, hydrogen gas may be vented from lead–acid batteries and this may form an explosive mixture with air if the concentration exceeds 3.8% by volume. Adequate ventilation and correct charging arrangements should always be observed.

- *Electrical* – Batteries contain a lot of stored energy. Under certain circumstances, this energy may be released very quickly and unexpectedly. This can happen when the terminals are short circuited, for example with an uninsulated metal spanner or screwdriver. When this happens, a large amount of electricity flows through the metal object, making it very hot, very quickly. If the battery explodes, the resulting shower of molten metal and plastic can cause serious burns and can ignite any explosive gases present around the battery. The short circuit may also produce ultraviolet (UV) light which can damage the eyes.

ASSESSMENT GUIDANCE

Battery voltage may be quite low but the current flow can be very high, especially under short-circuit conditions.

Most batteries produce quite low voltages, and so there is little risk of direct electric shock. However, as some large battery combinations produce more than 120 V d.c. precautions should be taken to protect users and those working in the vicinity of these installations.

- Ensure that live conductors are effectively insulated or protected.
- Keep metal tools and jewellery away from the battery.
- Post suitable notices and labels warning of the danger.
- Control access to areas where dangerous voltages are present.
- Provide effective ventilation to stop dangerous levels of hydrogen and air or oxygen accumulating in the charging area.

If battery cables are removed for any reason, ensure that they are clearly marked 'positive' and 'negative' so that they are reconnected with the correct polarity.

Minimising hazards from information and technology (IT) equipment

All modern offices contain computers that are operated as standalone units or networked (linked together). Most computer systems are sensitive to variations in the mains supply, which can be caused by external mains switching operations, faults on adjacent electricity distribution network circuits, which cause a voltage rise, and industrial processes such as induction furnaces and electric arc welding. These system disturbances are often well within the electricity supply companies' statutory voltage limits but may be sufficient to cause computers to crash.

To avoid the effects of this system 'noise', computer networks are normally provided with a clean supply, including a 'clean earth'. This is obtained by taking the supply to the computer network from a point as close as possible to the supply intake position of the building. The fitting of noise suppressors or noise filters provides additional protection to the computer system. If a production process disruption or data loss is a serious business risk, an uninterruptable power supply (UPS) is a sensible solution.

A UPS provides emergency power from a separate source (normally batteries) when power from the local supply company is unavailable. It differs from an auxiliary or emergency power system or standby generator, as these do not provide instant protection from a temporary power interruption. A UPS can be used to provide uninterrupted power to equipment, typically for 5 to 15 minutes, until an auxiliary power supply can be turned on, mains power is restored, or equipment is safely shut down. UPS units come in sizes ranging from units that will back up a single computer without monitor (around 200 W) to units that will power entire data centres or buildings (several megawatts, MW).

Small and large UPSs. Many single computers and installations have uninterruptable power supplies, so they may be live even if the mains power is off

All of the above features are designed to keep power supplies constant for IT and similar equipment to prevent data loss, but fault diagnosis can impact on this requirement. Before any fault diagnosis is carried out, confirm whether the standby or UPS will present a danger to the electrician undertaking the work. Remember, the preferred method of work is with the system dead and, as such, it must be isolated from *all* points of supply, including auxiliary circuits, and dual or alternative supplies, such as a UPS or a standby generator (that may be set for automatic start-up).

If it is necessary to isolate the computer supply network, permission must be sought from the duty holder or the responsible person for the computer systems to avoid any loss of data or damage to the computers. However, standby or a UPS backup systems can also be used to avoid IT shutdowns during fault diagnosis and testing by maintaining supplies to critical areas.

> **KEY POINT**
>
> Proving dead at the point of work is the most important part of ensuring a safe system of work. If the isolation has been incorrectly applied, the act of proving dead should identify that the circuit is, in fact, still live. A proper procedure, using a suitable proving device, is therefore essential (see page 147).

Minimising hazards from hazardous areas

Although all the hazards discussed above could be considered to give rise to hazardous areas, a 'hazardous area' is defined in the Dangerous Substances and Explosive Atmospheres Regulations 2002 (DSEAR) as 'any place in which an explosive atmosphere may occur in quantities such as to require special precautions to protect the safety of workers'.

These regulations provide a specific legal requirement to carry out a hazardous area study and document the conclusions in the form of zones. In the context of the definition, 'special precautions' relate to the construction, installation and use of apparatus, as given in BS EN 60079-10.

Hazardous areas are classified into zones based on an assessment of the frequency of the occurrence and duration of an explosive gas atmosphere, as follows:

- Zone 0: an area in which an explosive gas atmosphere is present continuously or for long periods

- Zone 1: an area in which an explosive gas atmosphere is likely to occur in normal operation

- Zone 2: an area in which an explosive gas atmosphere is not likely to occur in normal operation and, if it occurs, will only exist for a short time.

Most volatile materials (those that disperse readily in air) only form explosive mixtures between certain concentration limits. The flash point of a volatile material is the lowest temperature at which it can vaporise to form an ignitable mixture in air; flash point refers to both flammable liquids and combustible liquids.

The primary risk associated with combustible gases and vapours is the possibility of an explosion. Explosion, like fire, requires three elements: fuel, oxygen and an ignition source. Each combustible gas or vapour will ignite only within a specific range of fuel/oxygen mixtures.
Too little or too much gas and the gas or vapour will not ignite. These conditions are defined by the lower explosive limit (LEL) and the upper explosive limit (UEL). Any combustible gas falling between the two limits is potentially explosive.

Sources of ignition should be effectively controlled in all hazardous areas by a combination of design measures and systems of work. All electrical equipment must carry the appropriate markings if the integrity of the wiring system is to be maintained. A luminaire to IP 65 (this International Protection code means the fitting is dust-tight and water-jet protected) is an example of equipment used in hazardous areas.

When any work is to be undertaken on electrical equipment in a hazardous area, the following procedures must be observed:

- a written safe work procedure (PTW system) that includes control of activities which may cause sparks, hot surfaces or naked flames should be in place

- control of the working area with signs and barriers

- use of appropriate PPE

- only authorised and competent people to be engaged in the work activity

- only approved tools and equipment to be used

- prohibition of smoking, use of matches and lighters.

If work is to be undertaken in a hazardous area, the hazards and the requirements to be observed will normally be explained during the site induction process.

KEY POINT

Flash point is a term used to refer to liquids that are **flammable**.

Gases are **ignitable** and we use LEL and UEL to define limits between which a gas will ignite or explode.

Flammability is a measure of how easy it is for something to burn or ignite, causing a fire.

Combustion is a chemical reaction between fuel, oxygen and an ignition source (eg matches). The fuel can be a gas, a liquid or a solid such as wood or paper.

KEY POINT

An area in which an explosive mixture is present is called a **hazardous area** and precautions need to be taken when installing or working on any electrical equipment.

ACTIVITY

Can you name three appliances which may come under the heading of '**Hot work**'?

Hot work

Work of any type which involves actual or potential sources of ignition and which is done in an area where there may be a risk of fire or explosion (for example welding, flame cutting and grinding).

HAZARDS OF FAULT DIAGNOSIS

The Electricity at Work Regulations 1989 (EAWR) require those in control of part or all of an electrical system to ensure that it is safe to use and that it is maintained in a safe condition. The process of identifying and rectifying faults will inevitably involve exposure to some hazards and these must be dealt with in accordance with the requirements of the EAWR.

Electrical injuries are the most likely hazard and the following may occur due to contact with mains 230 V electricity:

- electric shock
- electrical burns
- loss of muscle control
- fires arising from electrical causes
- arcing and explosion.

Electric shock

Electric shock may arise either from direct contact with a live part during the testing process or, indirectly, by contact with an exposed conductive part that has become live as a result of a fault condition such as:

- contact with live electrical parts if basic protection has been removed
- a cracked equipment case causing 'tracking' from internal live parts to the external surface
- poor installation practice exposing bare live conductors at terminations
- exposure to static electricity from industrial processes or something as simple as walking on a carpet.

The magnitude (size) and duration of the shock current are the two most significant factors determining the severity of an electric shock.

The magnitude of the shock current depends on the contact voltage and impedance (electrical resistance) of the shock path.

Assessment criteria

4.1 State the dangers of electricity in relation to the nature of fault diagnosis work

ASSESSMENT GUIDANCE

Ensure you keep up to date with the latest treatments for electric shock. You may be asked to describe the actions you would take.

ACTIVITY

Using any source of reference material, find out why an electric cattle fence with an operating voltage of 8000 V does not cause electrocution.

ACTIVITY

You should have access to a poster showing the treatment to be used in the event of an electric shock. See if you can find one in your workshop or workplace.

A possible shock path always exists through ground contact (for example, through a hand-to-feet route); in this case the shock path impedance is the body impedance plus any external impedance. A more dangerous situation is a hand-to-hand shock path, when one hand is in contact with an exposed conductive part, such as an earthed metal equipment case, while the other simultaneously touches a live part. In this case, the current will be limited only by the body impedance.

As the voltage increases, so the body impedance decreases – this increases the shock current. When the voltage decreases the body impedance increases, which reduces the shock current.

Systems where voltages are below 50 V a.c. or 120 V d.c. (extra-low voltage) reduce the risk of electric shock to a low level. If the system energy levels are low, then arcing is unlikely to cause burns.

Electric burns

Electric burn injury may arise due to:

- the passage of shock current through the body, particularly if at high voltage
- exposure to high-frequency radiation, for example, from radio transmission antennas.

An electrical burn may not show on the skin at all or may appear minor, but the damage can extend deep into the tissues beneath the skin. Medical advice should always be sought if an electrical burn is sustained.

Loss of muscle control

People who experience an electric shock often get painful muscle spasms that are strong enough to break bones or dislocate joints. This loss of muscle control may mean that the person cannot let go to escape the electric shock. Alternatively, the person may fall if they are working at height or be thrown into nearby machinery and structures.

Fires arising from electrical causes

It is believed that in Britain each year there are over 30 000 fires in domestic and commercial premises that have electricity as a factor in their cause. The principal causes of fires arising from electricity are wiring with defects such as insulation failure, the overloading of conductors, lack of electrical protection, poor connections and the incorrect storage of flammable materials.

Arcing and explosion

This frequently occurs due to short-circuit flashover accidentally caused while working on live equipment (either intentionally or unintentionally). Arcing generates UV radiation, causing severe sunburn; molten metal particles are also likely to be ejected onto exposed skin surfaces.

There are two main electrical causes of explosion; short circuit, due to an equipment fault, and ignition of flammable vapours or liquids, caused by sparks or high surface temperatures.

VOLTAGE TESTING

Within commercial, industrial and some domestic installations, different supply voltages can be encountered. It is important that suitable methods (such as an instrument in accordance with GS 38) are used to verify these voltages.

Voltages that are likely to be encountered are:

- 400 V – industrial and commercial equipment, such as motors and other fixed plant
- 230 V – industrial, commercial and domestic, such as lighting and power supplies
- 110 V – construction sites have reduced low voltage system for tools and lighting
- SELV – Class III (no greater than 50 V), such as portable lamps.

Instruments used solely for voltage verification fall into two categories:

- test lamps, which rely on an illuminated lamp to show if a voltage is present
- voltage detector meters, which give a visual readout (analogue or digital) of the voltage.

It is recommended that test lamps and voltage indicators are clearly marked with the maximum voltage that may be tested and any short-time rating, if applicable. Knowledge of the expected voltage enables the correct choice of instrument to be made and ensures its maximum voltage is not exceeded.

Test lamps are fitted with a 15 W bulb (that may be protected by a guard) and should not give rise to danger if the lamp is broken. This type of detector requires overcurrent protection, which may be provided by a suitable high breaking capacity (hbc or hrc) fuse or fuses with a low current rating (usually not exceeding 500 mA), or by means of a current-limiting resistor and a fuse. These protective devices are housed in the probes themselves. The test lead or leads are held captive and sealed into the body of the voltage detector.

Assessment criteria

4.2 Describe how to identify supply voltages

SmartScreen Unit 308
Handout 20

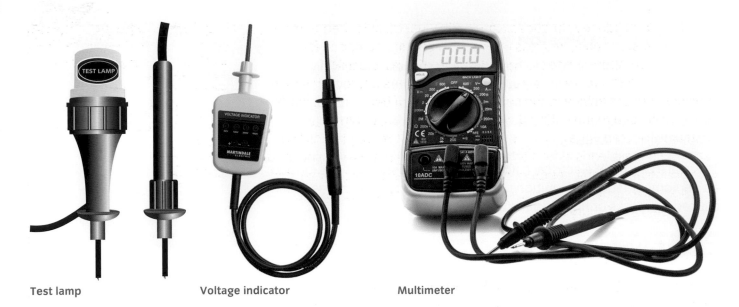

Test lamp   Voltage indicator   Multimeter

Voltage indicators/detectors usually incorporate two or more independent indicating systems – visual and audible – and limit energy input to the detector by means of internal circuitry. They are provided with in-built test features to check the functioning of the detector before and after use. If the detector does not have an in-built test feature it must be tested before and after application to the test area.

It is recommended that test lamps and voltage indicators are clearly marked with the maximum voltage that may be tested by the device and any short-time rating for the device, if applicable. This rating is the recommended maximum current for the device and the maximum time for which it should be sustained. These devices are generally not designed to be connected for more than a few seconds. Knowledge of the expected voltage will assist in making sure that the correct instrument is chosen and its maximum voltage is not exceeded.

USE OF INSTRUMENTS FOR FAULT DIAGNOSIS

In fault diagnosis, the use of suitable and safe voltage-indicating devices and measuring instruments is as important as the competency of the person undertaking the fault-finding activities. The possibility of touching live parts is increased during electrical testing and fault finding, when conductors at dangerous voltages are often exposed. The risks can be reduced if testing is done while the equipment or part of an installation is made dead and is isolated from any dangerous source of electrical supply. Special attention should be paid when carrying out tests with instruments capable of generating test voltages greater than 50 V or which use the supply voltage for the purpose of earth-loop testing or a residual-current device test. Refer back to page 146 for more information about electric shock.

The HSE has produced Guidance Note GS 38 (Electrical test equipment for use by electricians), which is intended to provide guidance for electrically competent people, including electricians, electrical contractors, test supervisors, technicians, managers and/or appliance retailers. It offers advice in the selection and use of test probes, leads, lamps, and voltage-indicating devices and measuring equipment for circuits with rated voltages not exceeding 650 V. It recommends the use of fused test leads aimed primarily at reducing the risks associated with arcing under fault conditions. Where possible, it is recommended that tests are carried out at reduced voltages, which will usually reduce the risk of injury.

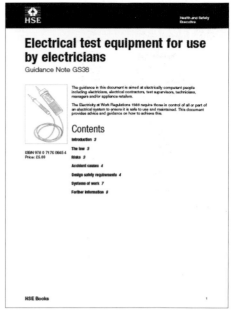

Guidance Note GS 38, an essential tool for electricians and others

Probes and leads

Guidance Note GS 38 recommends that probes and leads used in conjunction with voltmeters, multi-meters and electricians' test lamps or voltage indicators should be selected to prevent danger.

Probes should:

- have finger barriers or stops, or be shaped so that the hand or fingers cannot make contact with live conductors under test
- be insulated to leave an exposed metal tip not exceeding 4 mm measured across any surface of the tip. Where practicable, it is strongly recommended that this is reduced to 2 mm or less, or that spring-loaded retractable screened probes are used.

Leads should:

- be adequately insulated and coloured so that any one lead is readily identifiable from any other
- be flexible and sufficiently robust
- be long enough for the purpose – but not too long
- not have accessible exposed conductors, even if they become detached from the probe or from the instrument
- have suitably high breaking capacity (hbc), sometimes known as hrc, fuse or fuses with a low current rating (usually not exceeding 500 mA), or a current-limiting resistor and a fuse.

GS 38 also recommends that, if a test for the presence or absence of voltage is being made, the preferred instrument to be used is a proprietary test lamp or voltage indicator. The use of multi-meters for voltage indication has often resulted in accidents due to the multi-meter being set on the incorrect range.

ASSESSMENT GUIDANCE

When using a two-lead testing device, always place the first lead on the earth or neutral terminal. This will ensure that the free lead is not live.

KEY POINT

Voltage indicators and test lamps do not fail safe and, as such, must always be tested on a known voltage source before and after use.

Standards for test instruments

The basic instrument standard is BS EN 61557: 'Electrical safety in low voltage distribution systems up to 1000 V a.c. and 1500 V d.c. Equipment for testing, measuring or monitoring of protective measures'. This standard includes performance requirements and requires compliance with BS EN 61010: 'Safety requirements for electrical equipment for measurement control and laboratory use' which is the basic safety standard for electrical test instruments.

The following table shows the test instrument along with the associated harmonised standard.

Instrument	Standard	Instrument	Standard
insulation resistance ohmmeters	BS EN 61557-2	earth electrode resistance testers	BS EN 61557-5
earth fault loop impedance testers	BS EN 61557-3	RCD testers	BS EN 61557-6
low-resistance ohmmeters	BS EN 61557-4		

Instrument accuracy

A basic measuring accuracy of 5% is usually adequate for test instruments constructed in accordance with BS EN 61557 and for analogue instruments, a basic accuracy of 2% of full-scale deflection will provide the required accuracy measurement over the useful proportion of the scale.

Insulation resistance testers (BS EN 61557-2)

Insulation resistance test meters are used for insulation resistance testing and the separation of circuits, including SELV or PELV. Insulation tests should be made on electrically isolated circuits and, before commencing the test, it should be verified that pilot or indicator lamps and capacitors are disconnected from the circuits being tested, to avoid inaccurate test results. All voltage-sensitive electronic equipment, such as residual current circuit breakers (RCCBs), residual-current circuit breakers with overload protection (RCBOs), dimmer switches, touch switches, delay timers, power controllers, electronic starters for fluorescent lamps, emergency lighting and residual current devices (RCDs), should be disconnected so that they are not subjected to the test voltage. If these items cannot be disconnected, a measurement to protective earth with the line and neutral connected together should be made.

The operating accuracy of insulation resistance testers can be affected by 50 Hz currents induced into the cables under test and by capacitance in the test object. This capacitance may be as high as 5 µF. It is necessary for the test instrument to have a facility that discharges this capacitance safely. The test leads should not be touched while the tester is being used and, after the test, the instrument should be left connected until the capacitance within the installation has fully discharged.

Remember that the values shown when insulation resistance testing are in megohms (MΩ, millions of ohms).

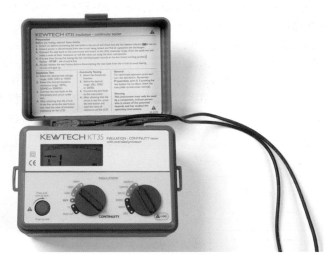

An insulation resistance tester

Finally, never measure resistance in a circuit when power is applied. If the circuit has sufficient power, the meter can explode or burst into flames. Most modern meters are fuse and diode protected, in order to prevent explosions, but would still be damaged by an overload of this magnitude.

Details of the method for completing insulation resistance tests can be found in Unit 307, pages 65–72.

Earth fault loop impedance tester (BS EN 61557–3)

These instruments operate by drawing a test current (25 A) from the supply and causing it to flow around the earth fault loop. Earth fault loop impedance is required to verify that there is an earth connection and that the value of the earth fault loop impedance is less than or equal to the value determined by the designer and which was used in the design calculations.

The earth fault current loop comprises the following elements, starting at the point of fault on the phase to earth loop:

- the protective conductor
- the main earthing terminal and earthing conductor
- for TN systems (TN-S and TN-C-S), the metallic return path or, in the case of TT systems, the earth return path
- the path through the earthed neutral point of the transformer
- the source phase winding
- the line conductor from the source to the point of fault.

ACTIVITY

If a meter is inadvertently connected to the mains supply it is reasonable to assume that some damage may have occurred. What course of action would you take?

ACTIVITY

Short circuits may often blow clear. What is meant by this?

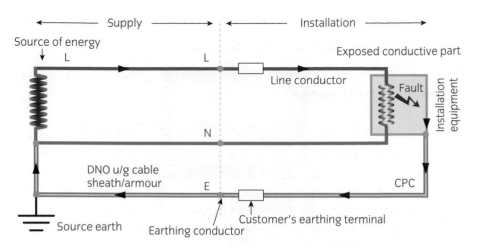

Earth fault loop path

Earth fault loop

Earth fault loop impedance testing is not recommended for fault diagnosis but can be used to confirm that any repair or rectification has not altered the maximum permissible earth fault loop impedances.

Details of the method for carrying out earth fault loop impedance tests can be found in Unit 307, pages 75–86.

Low-resistance ohmmeters (BS EN 61557-4)

These are used for checking protective conductor continuity, including main and supplementary equipotential bonding, ring final circuit continuity tests and polarity. Other applications include the checking of the integrity of welded joints, bolted lap joints on bus bars and verifying winding resistance in large motors and transformers. The test current may be d.c. or a.c. and should be derived from a source with a no-load voltage of not less than 4 V, and no greater than 24 V and a short-circuit current of not less than 200 mA.

The measuring range should span $0.2\ \Omega$ to $2\ \Omega$ with a resolution of at least $0.01\ \Omega$ for digital instruments and, for continuity testing, the lowest ohms scale must be selected – low values are expected.

Note that general-purpose multi-meters are not capable of supplying these voltage and current parameters.

Field effects contributing to in-service errors are contact resistance, test lead resistance, a.c. interference and thermocouple effects in mixed metal systems.

Earth electrode testers (BS EN 61557-2)

There are three methods of measuring the resistance of an earth electrode:

- E1 uses a dedicated earth electrode tester (fall-of-potential three- or four-terminal type)

Low resistance ohmmeter

- E2 uses a dedicated earth electrode tester (stakeless or probe type)
- E3 uses an earth fault loop impedance tester.

Earth electrode resistance testing will not be described in this unit. It is covered in Unit 307, page 81.

Residual current tester (RCD) (BS EN 61557-6)

RCDs are sensitive devices, typically operating on earth fault currents as low as 30 mA and with response times as fast as 20 to 40 ms. BS 7671 provides for wider use of RCDs, for indoor circuits as well as for socket outlets intended to supply portable outdoor electrical equipment that is outside the zone of earthed equipotential bonding.

Where an installation incorporates an RCD, Regulation 514.12.2 requires a notice to be fixed in a prominent position at or near the origin of the installation. The notice should be in indelible characters not smaller than those illustrated in the Regulation.

Note that the integral test button incorporated in RCDs only verifies the correct operation of the mechanical parts of the RCD and does not provide a means of checking the continuity of the earthing conductor, the earth electrode or the sensitivity of the device. This can only be done by use of an RCD tester designed for testing RCDs.

RCD testers should be capable of applying the full range of test current to an in-service accuracy, as given in BS EN 61557-6 (10%). This in-service reading accuracy will include the effects of voltage variations around the nominal voltage of the tester. To check the RCD operation and to minimise danger during the test, the test current should be applied for no longer than 2 seconds.

Details of the method for testing the correct operation of residual current devices are shown on page 103 in Unit 307.

Measuring the prospective fault current, I_{pf}

The current that is expected to flow from line to neutral in the event of a fault is the prospective short-circuit current (I_p or PSCC) and from line to earth it is the prospective earth-fault current (I_f or PEFC).

The fault current depends on the total circuit impedance at the point of the fault. If the value of either the prospective short-circuit current or the prospective earth-fault current is higher than that which can be safely interrupted by the protective devices, catastrophic damage to the protective devices or associated equipment may occur. Typically, such fault currents are measured in kA. Regulation 612.11 requires both the prospective short-circuit current and the prospective earth-fault current to be measured, calculated or determined by another method, at the origin and at every other relevant point in the installation.

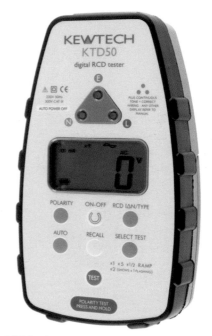

RCD tester

The reference to 'every other relevant point' means every point where a protective device is required to operate under fault conditions and this includes the origin of the installation.

Prospective fault levels should be measured at the distribution board, with all main bonding in place, between live conductors and between line conductors and earth.

To measure the prospective fault current, the prospective fault-current range of a suitable earth fault loop impedance tester can be used (BS EN 61557-3). Two-lead and three-lead testers are available for measuring prospective fault current. It is important that any instrument being used is set on the prospective fault current range and is connected in accordance with the manufacturer's instructions. Note that instrument accuracy decreases as scale reading increases.

Remember, this is a live test and, as the tester will be using the mains supply, care must be taken not to touch exposed terminals or other exposed metalwork during testing. The guidance contained in GS 38 for the test probes and leads should be followed.

Details of the method for checking prospective fault current can be found on page 99 of Unit 307.

Tong tester or clamp on ammeter

It may be necessary to measure current draw within a circuit to establish whether a circuit or entire installation is overloading. Traditional ammeters need to be connected in series with the load which requires breaking the circuit cable. As this is both impractical and dangerous, clamp on ammeters can measure current carried by a conductor by clamping it around the cable. The instrument will determine the current flow within the conductor by measuring the magnetic field strength around the conductor. As the field strength is proportional to current flow, an accurate measure of current can be detected safely and without disconnection or disruption.

Phase sequence tester

Phase sequence test instruments detect the rotation of a three-phase supply. As motors/machines rely on correct phase sequence for correct direction of rotation, it is essential that phase sequence is checked.

USING INSTRUMENTS SAFELY

When using test instruments to carry out fault diagnosis, follow these basic precautions to achieve safe working.

- *Understanding the equipment* – make sure you are familiar with the instrument to be used and its ranges; check its suitability for the characteristics of the installation it will be used on.

- *Self test* – many organisations regularly test instruments on known values to ensure they remain accurate. These tests would be documented.

- *Calibration* – all electrical test instruments should be calibrated on a regular basis. The time between calibrations will depend on the amount of usage that the instrument receives, although this should not exceed 12 months under any circumstances. Instruments have to be calibrated under laboratory conditions, against standards that can be traced back to national standards; therefore, this usually means returning the instrument to a specialist laboratory. Once calibrated, the instrument will have a calibration label attached to it stating the date the calibration took place and the date the next calibration is due. It will also be issued with a calibration certificate, detailing the tests that have been carried out, and a reference to the equipment used. Instruments that are subject to any electrical or mechanical misuse (for example, if the instrument undergoes an electrical short circuit or is dropped) should be returned for recalibration before being used again. Electrical test instruments are relatively delicate and expensive items of equipment and should be handled with care. When not in use, they should be stored in clean, dry conditions at normal room temperature. Care should also be taken of instrument leads and probes, to prevent damage to their insulation and to maintain them in safe working condition.

- *Check test leads* – make sure that these and any associated probes or clips are in good order, are clean and have no cracked or broken insulation. Where appropriate, the requirements of the HSE Guidance Note GS 38 should be observed for test leads.

- *Select appropriate scales and settings* – it is essential that the correct scale and settings are selected for an instrument. Manufacturers' instructions must be observed under all circumstances.

HOW TO DOCUMENT FAULT DIAGNOSIS

BS 7671 requires that, following the inspection and testing of all new installations, alterations and additions to existing installations or periodic inspections, an Electrical Installation Certificate (EIC), together with a Schedule of Test Results, should be given to the person ordering the work; this is normally the client, duty holder or responsible person. Model forms for certification and reporting are contained in Appendix 6 of BS 7671.

Assessment criteria

4.4 Describe how to confirm test instruments are fit for purpose, functioning correctly and are correctly calibrated

ASSESSMENT GUIDANCE

You will be expected to be seen checking the instruments before and after use.

KEY POINT

The user of the instrument should always check to ensure that the instrument is within the calibration period before using it.

Assessment criteria

4.5 State the appropriate documentation that is required for fault diagnosis work and explain how and when it should be completed

There are two options available for the EIC, Form 1 or Form 2.

- *Form 1:* short form EIC, is to be used when one person is responsible for the design, construction, inspection and testing of an installation.
- *Form 2:* this EIC is to be used when more than one person is responsible for the design, construction, inspection and testing of an installation.

Whichever EIC is used, appropriate numbers of the following forms are required to accompany the certificate:

- Schedule of Inspections
- Schedule of Test Results.

When an addition to an electrical installation does not involve the installation of a new circuit a Minor Electrical Installation Works Certificate (MEIWC) may be used. This certificate is intended for use when work such as the addition of a socket outlet or lighting point to an existing circuit or a repair or modification to a circuit is undertaken.

EICs, EICRs and MEIWCs must be completed and signed by a competent person or persons in respect of the design, the construction and the inspection and test of the installation.

A competent person is defined in BS 7671 as: 'a person who possess sufficient technical knowledge, relevant practical skills and experience for the nature of the electrical work undertaken and is able at all times to prevent danger and, where appropriate, injury to him/herself and others'.

Therefore, competent persons must have a sound knowledge and relevant experience of the type of work being undertaken and of the technical requirements of BS 7671. They must also have a sound knowledge of the inspection and testing procedures contained in the Regulations and must use suitable testing equipment.

EICs and MEIWCs must identify who is responsible for the design, construction, inspection and testing, whether this is new work or an alteration or addition to an existing installation.

Form 1 (short form) Electrical Installation Certificate

Form 1, the short form EIC is a shortened version of Form 2. It is used where one person is responsible for the design, construction, inspection and testing of a new installation or where a major alteration or addition to an existing installation is carried out and the installation, inspection and testing of that work is the responsibility of one person.

Form 2 Electrical Installation Certificate (three signatory version)

Form 2, the EIC (three-signatory version) is used where more than one person is responsible for the design, construction, inspection and testing of a new installation or where a major alteration or addition to an existing installation is carried out. The Form 2 certificate also has space for the name, address, telephone number and signatures of two different design organisations. This can be used where different sections of the installation are designed by different individuals or companies (for example, the electrical installation may be designed by one consultant and the fire alarm system by another). In this situation the details and signatures of both parties are required. BS 7671 provides advice notes for the completion of each type of form and also guidance notes for the person receiving the certificate.

Electrical Installation Condition Report

The EICR should only be used for reporting on the condition of an electrical installation. It is intended primarily for the person who is ordering the work to be undertaken (the client, duty holder or responsible person for the installation) and anyone subsequently involved in additional or remedial work or additional inspections, to confirm, so far as reasonably practicable, whether or not the electrical installation is in a satisfactory condition for continued service.

Each observation of a problem or concern relating to the safety of an installation should be given an appropriate classification code selected from standard classification codes as follows:

- *C1: danger present*. Risk of injury. Immediate remedial action required.
- *C2: potentially dangerous*. Urgent remedial action required.
- *C3: improvement recommended*.

Electrical Installation Minor Works Certificate

The MEIWC is used for minor works, which are defined as an addition to an electrical installation that does not extend to the installation of a new circuit. This could cover, perhaps, the addition of a new socket outlet or of a lighting point to an existing circuit. This certificate includes space for the recording of essential test results but does not require the addition of a Schedule of Test Results.

Each certificate or report will have attached to it a set of guidance notes explaining its purpose.

When an electrical fault has been diagnosed, identified and then repaired the IET Wiring Regulations require that the circuit, system or individual piece of equipment be inspected, tested and functional tests carried out. These inspections and tests must be carried out in accordance with Part 6 of the Regulations and the results recorded in accordance with Part 6, Chapter 63. The requirements that apply to fault diagnosis and repair are:

- 631.2 – completion of an EICR
- 631.4 – the EICR shall be compiled and signed by a competent person or persons
- 631.5 – the report can be produced in any durable medium such as written hard copy or by electronic means.

Depending on the extent of the fault and subsequent report, either an EIC or a MEIWC will be issued to the person requesting the work.

The original certificate should be retained in a safe place and be shown to any person inspecting the installation or carrying out further work in the future. If the property is later vacated, this certificate will demonstrate to the new owner or occupier that the minor works covered by the certificate met the requirements of the Regulations at the time that the report was issued.

IMPLICATIONS OF FAULT DIAGNOSIS

Depending on the type of fault diagnosed and the method of rectification required, there may be a number of implications for those with responsibility for the electrical installation and users of the system.

There may be a requirement to disturb the fabric or structure of the building and, if this is the case, it is very important for all aspects of the rectification to be discussed with the client. Agreement must be obtained for the work to be undertaken, for the extent of the repair necessary (to brick, block, plaster, concrete, screed, plasterboard and decorations, for example) and for the contractual arrangements (who is paying for the repair).

Once agreement has been reached on the contractual issues, the repairs should be carried out in a logical manner, while taking into account the effect that circuit disconnections will have on users of the system. When only a small area or limited amount of equipment is affected by a fault, it is preferable not to request a total shutdown of the whole installation. By analysing circuit and schematic diagrams, it should be possible to isolate individual circuits to avoid disruption. If it is necessary to isolate individual circuits, especially those that supply IT equipment, then permission must be sought from the duty holder or the responsible person for the equipment, to avoid any loss or damage to such items as computers.

Assessment criteria

4.6 Explain why carrying out fault diagnosis work can have implications for customers and clients

ACTIVITY

Like any other installation work, maintenance work will cause disturbance to the work area and create a mess on the floor, carpets etc. Make a list of items required to keep the area clean and tidy.

ASSESSMENT GUIDANCE

Any temporary repairs made while awaiting delivery of replacement parts must be carried out to the same regulatory requirements as permanent installations.

There may, however, be circumstances when minor disruption or even a 'risk of trip' are not acceptable and alternative arrangements, such as 'out of hours working', will need to be considered.

USING TESTING INSTRUMENTS FOR FAULT DIAGNOSIS

Standards for test instruments

The basic instrument standard is BS EN 61557: 'Electrical safety in low voltage distribution systems up to 1000 V a.c. and 1500 V d.c. Equipment for testing, measuring or monitoring of protective measures'. This standard includes performance requirements and requires compliance with BS EN 61010: 'Safety requirements for electrical equipment for measurement control and laboratory use' which is the basic safety standard for electrical test instruments.

The requirements for the safe and correct use of instruments to be used for testing and commissioning has been dealt with in detail in Unit 307. However, the following sections give guidance on how test instruments may be used for completing fault diagnosis work.

Insulation resistance testers (BS EN 61557-2)

Insulation resistance testers are used to verify that there is adequate insulation resistance between live conductors, and also between live conductors and the protective conductor connected to earth. The test will verify that insulation is satisfactory to withstand the supply voltage and that live conductors or protective conductors are not short-circuited. A low insulation resistance reading could indicate damaged insulation, for example, a nail in a cable, dampness in a fitting, a faulty accessory or a faulty fitting.

Before use, the condition of the instrument and leads should be checked and confirmation made that it operates correctly (in open circuit and short circuit).

In an overload condition, the circuit conductors are carrying more current that the manufacturer's design specification. If this is allowed to continue indefinitely the conductors will become very hot, causing deterioration of the electrical insulation surrounding the live conductors. This may eventually lead to breakdown of the insulation and a short circuit. It is likely, however, that the protective device, if specified correctly, will operate before a short circuit condition is reached and so prevent an insulation failure occurring. Continual operation of a protective device should be investigated immediately.

Assessment criteria

4.7 Specify and undertake the procedures for carrying out the following tests and their relationship to fault diagnosis:
- continuity
- insulation resistance
- polarity
- earth fault loop impedance
- RCD operation
- current and voltage measurement
- phase sequence

ACTIVITY

What would be the typical output current of an insulation resistance tester with an output of 500 V d.c.?

ASSESSMENT GUIDANCE

When IR testing two-way or intermediate circuits, remember to operate the switches to include all cables in the circuit.

If it is suspected that insulation is beginning to fail, indicated by the frequent operation of the protective device, or even the operation of a RCD, the fault can be can be located by undertaking an insulation resistance test on the whole installation (if it is a relatively small system) or, if it is a complex installation, on the various component parts. If it is established that a fault condition exists when testing at the tails to the distribution board, the next step is to check each circuit in turn.

Where the fault involves the line conductor(s), testing must be done with only one fuse inserted at a time or with one circuit breaker closed at a time. For a neutral fault, neutral conductors must be removed from the neutral bar in order to identify the fault circuit.

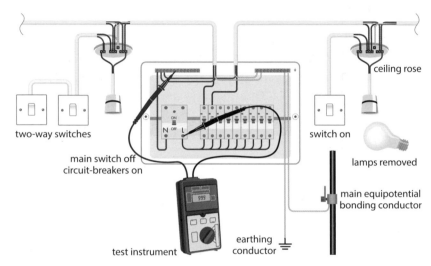

Insulation resistance test of a whole consumer unit

The test is carried out between the live (line and neutral) conductors and the circuit protective conductors at the distribution board. (All equipment should be removed before the test is carried out.) If a low reading is recorded, a socket at the centre of the ring should be selected and the conductors at the socket disconnected. Carry out an insulation test in both directions. It is likely that one side of the ring will indicate a fault. If so, subdivide the section being worked on – test and subdivide until the faulty section of cable or fitting is identified.

When trying to identify a fault, all circuit wiring should be included in the test – all switches must be closed and all current-using equipment, such as lamps and fixed loads, must be disconnected.

Remember, if functional switching is likely to exclude part of the circuit from the test, the switches must be operated during the test. For example, two-way switching in a lighting circuit should be operated one switch at a time to ensure that both strappers and the switch have all been tested.

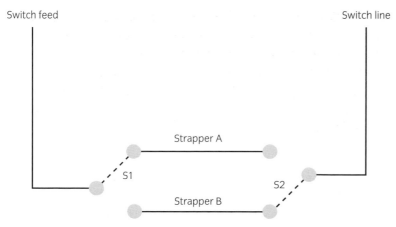

Switch feed

Switch line

Strapper A

S1

S2

Strapper B

ASSESSMENT GUIDANCE

The same rule applies to the two-way 'converted' arrangement.

If the test is carried out with switches as shown, Strapper B and switch line will not be tested. Operating switch S2 and retesting will test switch line but not Strapper B. To include Strapper B, switch S1 must be operated as well.

Two-way switching

Low-voltage installations operating at 230/400 V must be tested at 500 V d.c. Where surge protection devices may influence the test results and it is not practicable to disconnect them, the test may be conducted at 250 V but a minimum 1.0 MΩ is required.

Insulation resistance tests should be carried out on electrically isolated circuits and, before commencing the test, it should be verified that pilot or indicator lamps and capacitors are disconnected from the circuits being tested (to avoid inaccurate test values being obtained). All voltage-sensitive electronic equipment, such as RCCBs, RCBOs, dimmer switches, touch switches, delay times, power controllers, electronic starters for fluorescent lamps, emergency lighting and residual current devices (RCDs), should be disconnected so that they are not subjected to the test voltage. If these items cannot be disconnected, a measurement to protective earth, with the line and neutral connected together, should be made.

For three-phase installations, similar tests should be conducted, with a test being carried out between all 10 pairs of conductors.

Remember when insulation resistance testing that the values shown are in megohms (MΩ, millions of ohms).

Finally, never measure resistance in a circuit when power is applied. If the circuit has sufficient power, the meter can explode or burst into flames. Most modern meters are fuse and diode protected in order to prevent explosions, but would still be damaged by an overload of this magnitude.

Low resistance ohmmeters (BS EN 61557-4)

These are used for checking protective conductor continuity, including main and supplementary equipotential bonding, ring final circuit continuity tests and polarity. Other applications include the checking

Low resistance ohmmeter

of the integrity of welded joints, bolted lap joints on bus bars and verifying winding resistance in large motors and transformers. The test current may be d.c. or a.c. and should be derived from a source with a no-load voltage of not less than 4 V and no greater than 24 V and a short-circuit current of not less than 200 mA.

Continuity testing may be used to locate an open circuit conductor. To locate the conductor, the circuit may need to be broken into sections. To identify a lighting circuit that is not operating correctly, close the switch and test for continuity at the various light positions, operating the switches to confirm switch circuits are intact.

The easiest procedure to follow with any fault is to split the circuit into sections and systematically to test each section until the fault is identified. On a ring final circuit, the ring should be split at the distribution board and again at the approximate mid-point in the ring. You should then test each of the two sections before splitting each leg again and again, testing each length of cable between accessories until the fault is identified.

A continuity test can also be used to verify the continuity of each conductor including the circuit protective conductors. The test can ensure that there are no cross connections in a ring final circuit that might cause a shortened ring with several socket-outlets fed from spurs.

The test is a three-step process requiring:

- measurement of the end-to-end resistance of each conductor at the origin of the ring
- with the line and neutral conductors cross connected, measurement between line and neutral at each socket outlet (if the cross connections are made incorrectly, the readings obtained will tend to rise in value towards the mid-point of the ring and fall towards the origin)
- with the line and circuit protective conductors cross connected, measurement between line and cpc at each socket outlet (provided the cross connections are made correctly, all readings should be approximately equal to one quarter that of the first step, with the exception of spurs).

Locating a fault on a radial circuit for a single item of fixed equipment may be undertaken by disconnecting all of the conductors at the consumer unit or distribution board and linking them together in a connector block.

Using a low resistance ohmmeter, test back to the consumer unit for continuity between line, neutral and the cpc. If there is a break in the circuit and no apparent branches, there could be a concealed junction box within the wiring. This junction box must be located and further tests undertaken to ensure continuity in both directions.

ACTIVITY

Describe how a half-test method could be applied to a ring final circuit.

Residual current tester (RCD) (BS EN 61557-6)

RCDs are sensitive devices, typically operating on earth fault currents as low as 30 mA and with response times as fast as 20 to 40 ms. BS 7671: 2008 provides for wider use of RCDs, for indoor circuits as well as for socket outlets intended to supply portable outdoor electrical equipment that is outside the zone of earthed equipotential bonding.

If an insulation resistance test has been carried out on a suspected faulty circuit with a 30 mA RCD, supplying an outside light and there is no low insulation resistance reading or other apparent fault, check the correct functioning of the RCD by injecting ½ × rated current and 1 × rated current. If the RCD trips at 15 mA, test its sensitivity by means of a ramp test.

Most modern RCD testers include a ramp test facility. The ramp test slowly increases the tripping current until the RCD trips.

The tester display is in milliamps (mA) and readings will identify if an RCD is susceptible to nuisance tripping or is out of specification.

The equipment can also be used to indicate which circuits have high leakage currents and can be used when there is evidence of frequent trips on a particular RCD.

The installation provides the source for RCD tests and, under certain circumstances, these tests can result in potentially dangerous voltages appearing on exposed and extraneous conductive parts within the installation. Therefore, suitable precautions must be taken to prevent contact with any such part.

Other users of the building should be made aware that tests are being carried out and warning notices should be posted, as necessary.

Voltage testing

Within commercial, industrial and some domestic installations, different supply voltages can be encountered. It is important that suitable methods (such as an instrument in accordance with GS 38) are used to verify these voltages.

Voltages that are likely to be encountered are:

- 400 V – industrial and commercial equipment, such as motors and other fixed plant
- 230 V – industrial, commercial and domestic, such as lighting and power supplies
- 110 V – construction sites have a reduced low voltage system for tools and lighting

- SELV – Class III (no greater than 50 V), such as portable lamps.

Instruments used solely for voltage verification fall into two categories:

- test lamps, which rely on an illuminated lamp to show if a voltage is present
- voltage detector meters, which give a visual readout (analogue or digital) of the voltage.

It is recommended that test lamps and voltage indicators are clearly marked with the maximum voltage that may be tested and any short-time rating, if applicable. Knowledge of the expected voltage enables the correct choice of instrument to be made and ensures its maximum voltage is not exceeded.

Test lamps are fitted with a 15 W bulb (that may be protected by a guard) and should not give rise to danger if the lamp is broken. This type of detector requires overcurrent protection, which may be provided by a suitable high breaking capacity (hbc or hrc) fuse or fuses with a low current rating (usually not exceeding 500 mA), or by means of a current-limiting resistor and a fuse. These protective devices are housed in the probes themselves. The test lead or leads are held captive and sealed into the body of the voltage detector.

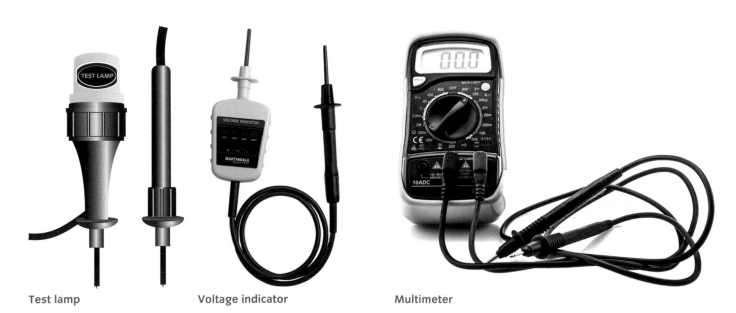

Test lamp   Voltage indicator   Multimeter

Voltage indicators/detectors usually incorporate two or more independent indicating systems – visual and audible – and limit energy input to the detector by means of the internal circuitry. They are provided with in-built test features to check the functioning of the detector before and after use. If the detector does not have an in-built test feature it must be tested before and after application to the test area.

WHAT TO DO NEXT?

Following an inspection and test on an electrical installation, an Electrical Installation Condition Report will be provided to highlight any safety shortcomings, defects or deviations from the current version of BS 7671.

An Electrical Installation Condition Report is intended primarily for the person who is ordering the work to be undertaken (the client, duty holder or responsible person for the installation) and anyone subsequently involved in additional or remedial work or additional inspections, to confirm, so far as reasonably practicable, whether the electrical installation is in a satisfactory condition for continued service.

Each observation relating to a problem or safety concern should be given an appropriate classification code selected from standard classification codes.

Assessment criteria

4.8 Identify whether test results are acceptable and state the actions to take where unsatisfactory results are obtained

Code C1: danger present

This code signifies a risk of injury and that immediate remedial action is required.

Where practicable, items classified as C1 should be made safe on discovery by the person undertaking the inspection. The duty holder or responsible person ordering the report should be advised of the action taken and asked for permission to allow remedial work to be undertaken straight away. If that is not practical, other appropriate action should be taken, such as switching off and isolating the affected parts of the installation to prevent danger.

Code C2: potentially dangerous

This code signifies that urgent remedial action is required.

For items classified as C2, the safety of those using the installation may be at risk and it is recommended that a competent person undertakes any necessary remedial work as a matter of urgency.

Code C3: improvement recommended

For items classified as C3, the inspection has revealed an apparent deficiency which could not, due to the extent or limitations of the inspection, be fully identified.

It should be reported to the person requesting the inspection that further investigation is required, as soon as possible ,to determine the nature and extent of the apparent deficiency. For safety reasons, the electrical installation should be reinspected at appropriate intervals by a competent person.

The recommended date by which the next inspection is due should be included in the report, under 'Recommendations' and on a label near to the consumer unit or distribution board.

On completion of the inspection, the client, duty holder or responsible person ordering the report should receive the original and the inspector should retain a duplicate copy. The original report should be retained in a safe place and be made available to any person inspecting or undertaking work on the electrical installation in the future. If the property is vacated, the report will provide the new owner or occupier with details of the condition of the electrical installation at the time the report was issued.

Failure to take action when a deviation from BS 7671 has been identified and recorded would be seen as a failure under the provisions of Health and Safety at Work etc Act (1974) and the Electricity at Work Regulations 1989, and directors and managers of any company that employs more than five employees can be held personally responsible for failure to control health and safety.

OUTCOME 5

Understand the procedures and techniques for correcting electrical faults

FACTORS THAT AFFECT FAULT RECTIFICATION

Some faults are minor and these can be dealt with quickly and right after the fault has been diagnosed, with little impact on materials and manpower. However, in the case of major faults in large industrial or commercial enterprises, the repair time could be weeks or months, depending on the extent of the failure. If this is the case, careful planning of the work process is required and consideration should be made of whether the likely value of the repair work will necessitate a tendering process.

Agreeing the scope of the work through a tendering process

The next stage following a tender is the preparation of a contract document between the successful tenderer and the client, to ensure that all parties are aware of their obligations. Failure to prepare a contract could result in future conflict over costs, exact work requirements and the procedure for dealing with variations from the agreed job specification.

A contract of this nature usually includes:

- scope of the work to be done
- location of the work
- the provision of information, such as drawings, plans, certificates
- details of the payment schedule – a fixed price, or time and materials arrangement
- the completion dates and schedules
- details of responsibilities that may impact on the schedules, including those of any third parties
- responsibility for additional fees or charges that may arise over the course of the project
- responsibilities and requirement for safety of the contractor's staff and others who may be affected by the work

Assessment criteria

5.1 Identify and explain factors which can affect fault correction, repair or replacement

SmartScreen Unit 308
Handout 22 and Worksheet 22

ASSESSMENT GUIDANCE

It is not always possible to provide an exact quotation for the cost — an estimate is more likely. The customer must be kept aware of any developments.

- responsibility for losses, material damage or personal injuries that may occur during the project
- agreed variation procedure
- guarantee or warranty period for the completed work.

Downtime and costs are probably the two issues of greatest concern that must be agreed and then adhered to. Most clients prefer to fix a price and will not countenance additional costs, unless it can be proven that these result from a departure from the original specification, that could not have been foreseen.

When a quote for work is accepted, it is essential that the quoted work is undertaken; any deviation from the quote, such as not using specified components or using different-size cabling, could be construed as a breach of contract.

Procurement of a backup supply

Depending on the nature of the business affected by the fault, the client may request out-of-hours fault repair work. If this is not an option, the provision of temporary standby supplies (in the form of a backup generator) may be requested. The procurement and provision of suitable standby supplies could have an effect on the timescale for the fault rectification. Again, the supply and management of such equipment must be agreed before it is connected to the installation.

Other factors that affect timescale

Even with all agreements and arrangements in place, remember that, before the repair work can take place, an approved safe working procedure including a risk assessment must be prepared.

Any delay in providing information, such as past inspection and tests results, schematics or wiring diagrams, may have a bearing on how quickly the repair can be actioned.

Other factors may also influence the timescale. For example, if the fault has affected a functioning oven or furnace, some considerable time may need to elapse to allow the oven or furnace to cool sufficiently to allow personnel to enter the area.

Assessment criteria

5.2 Specify the procedures for functional testing and identify tests that can verify fault correction

TESTING THE CIRCUIT

When the repair work is complete, the circuit, system or individual piece of equipment must be inspected, tested and functional checks carried out in accordance with BS 7671 Chapter 61. This inspection and function testing will confirm the electrical integrity of the system before it is energised.

The tests recommended in Regulation 612, Testing of BS 7671, are divided into those tests that are conducted with the system dead and those that are done with the system live.

Tests conducted before the supply is reconnected include:

- continuity of protective conductors, including main and supplementary bonding
- continuity of ring final circuit conductors, including protective conductors
- insulation resistance.

Tests conducted with the supply connected are:

- to check the polarity of the supply, using an approved voltage tester
- earth electrode resistance, using a loop impedance tester
- earth fault loop impedance
- prospective fault current measurement
- functional testing, including RCDs and switchgear.

For detailed information on how the above testing is carried out, refer to Unit 307.

USING THE CORRECT FORMS

Once faults have been rectified, depending on the extent of the fault, repair work will require certification. There are two options available for the Electrical Installation Certificate (EIC): Form 1 or Form 2.

- *Form 1:* short form EIC, is to be used when one person is responsible for the design, construction, inspection and testing of an installation. An example of this form is shown on page 212.
- *Form 2:* this EIC is to be used when more than one person is responsible for the design, construction, inspection and testing of an installation. An example of this form is shown on page 213.

Whichever EIC is used, appropriate numbers of the following forms are required to accompany the certificate:

- Schedule of Inspections
- Schedule of Test Results.

When an addition to an electrical installation does not involve the installation of a new circuit a Minor Electrical Installation Works Certificate (MEIWC) may be used. This certificate is intended for use when work such as the addition of a socket outlet or lighting point to an existing circuit or a repair or modification to a circuit is undertaken.

Electrical Installation Certificates and Minor Electrical Installation Works Certificates must be completed and signed by a competent person or persons in respect of the design, the construction and the inspection and test of the installation.

ASSESSMENT GUIDANCE

Make sure you select the correct form for the purpose intended. Make sure you use the correct terminology when identifying the form. It is not an inspection schedule, it is a Schedule of Inspections; it is not a test certificate but a Schedule of Test Results etc.

Form 1

Form No: 505513........./1

ELECTRICAL INSTALLATION CERTIFICATE
(REQUIREMENTS FOR ELECTRICAL INSTALLATIONS - BS 7671 (IET WIRING REGULATIONS))

DETAILS OF THE CLIENT   Mr T Brown
32 South St
Anytown, Surrey   Post Code: TO1 1ZZ

INSTALLATION ADDRESS   The Coffee Bean
31 Station Road
Anytown, Surrey   Post Code: TO3 2YF

DESCRIPTION AND EXTENT OF THE INSTALLATION Tick boxes as appropriate

Description of installation:
Re-wire of ground floor, on change of use.

New installation ☑

Addition to an existing installation ☐

Alteration to an existing installation ☐

Extent of installation covered by this Certificate:
Complete electrical re-wire of refurbished premises, on change of use from offices to cafe/snack bar.

(Use continuation sheet if necessary)   see continuation sheet No: ......

FOR DESIGN, CONSTRUCTION, INSPECTION & TESTING
I being the person responsible for the design, construction, inspection & testing of the electrical installation (as indicated by my signature below), particulars of which are described above, having exercised reasonable skill and care when carrying out the design, construction, inspection & testing hereby CERTIFY that the said work for which I have been responsible is to the best of my knowledge and belief in accordance with BS 7671:2008, amended to 2011 ...... (date) except for the departures, if any, detailed as follows:

Details of departures from BS 7671 (Regulations 120.3 and 133.5):
None

The extent of liability of the signatory is limited to the work described above as the subject of this Certificate.

Signature: *W Hastings*   Date: 21-Jan-2011   Name (IN BLOCK LETTERS): W HASTINGS

Company: Hastings Electrical
Address: 21 The Arches
Anytown, Surrey   Postcode: TO2 9YY   Tel No: 01022 999999

NEXT INSPECTION
I recommend that this installation is further inspected and tested after an interval of not more than ...5..... years/months.

SUPPLY CHARACTERISTICS AND EARTHING ARRANGEMENTS Tick boxes and enter details as appropriate

Earthing arrangements	Number and Type of Live Conductors	Nature of Supply Parameters	Supply Protective Device Characteristics
TN-C ☐	a.c. ☑ d.c. ☐	Nominal voltage, U/U₀ (1) ....230.. V	Type BS 1361 Fuse
TN-S ☐	1-phase, 2-wire ☑ 2-wire ☐	Nominal frequency, f (1) ....50. Hz	
TN-C-S ☑	1-phase, 3-wire ☐ 3-wire ☐	Prospective fault current, I_pf (2) ...9.0. kA	Rated current ....100.. A
TT ☐	2-phase, 3-wire ☐ other ☐	External loop impedance, Z_e (2) 0.28 ☒	
IT ☐	3-phase, 3-wire ☐		
Other sources ☐ of supply (to be detailed on attached schedules)	3-phase, 4-wire ☐	*(Note: (1) by enquiry, (2) by enquiry or by measurement)*	
	Confirmation of supply polarity ☑		

Form 1

Form No: 505513........../1

PARTICULARS OF INSTALLATION REFERRED TO IN THE CERTIFICATE Tick boxes and enter details, as appropriate

Means of Earthing

Distributor's facility ☑

Installation earth electrode ☐

Maximum Demand

Maximum demand (load) ...80. ~~kVA~~ / Amps *Delete as appropriate*

Details of Installation Earth Electrode *(where applicable)*

	Type (e.g. rod(s) tape etc)	Location	Electrode resistance to Earth
	N/A	N/A	N/A ☒

Main Protective Conductors

Earthing conductor: material Copper   csa ....16....mm²   Continuity and connection verified ☑

Main protective bonding conductors   material Copper   csa ....10....mm²   Continuity and connection verified ☑

To incoming water and/or gas service ☑   To other elements: N/A.

Main Switch or Circuit-breaker

BS, Type and No. of poles BS EN 60947-3 (2- pole)   Current rating ...100. A   Voltage rating ....230....V

Location Services cupboard adjacent rear exit   Fuse rating or setting........N/A. A

Rated residual operating current I_Δn = ...N/A....mA, and operating time of ....N/A.....ms (at I_Δn) *(applicable only where an RCD is suitable and is used as a main circuit-breaker)*

COMMENTS ON EXISTING INSTALLATION (in the case of an addition or alteration see Section 633):
Not Applicable.

SCHEDULES
The attached Schedules are part of this document and this Certificate is valid only when they are attached to it.
....1.... Schedules of Inspections and ....1.... Schedules of Test Results are attached.
(Enter quantities of schedules attached).

Form 1: Short form of Electrical Installation Certificate (always check you are using the latest forms, as found on the IET website: http://electrical.theiet.org)

Form 2 — Page 1

Form 2
Form No: SSSS13...../2

ELECTRICAL INSTALLATION CERTIFICATE
(REQUIREMENTS FOR ELECTRICAL INSTALLATIONS - BS 7671 [IET WIRING REGULATIONS])

DETAILS OF THE CLIENT
Mr D Roberts
23 Acacia Avenue,
Sandtown, Berks.   Post Code: S10 0JT

INSTALLATION ADDRESS   Unit 3 The Quadrant
Sandtown, Business Park,
Sandtown, Berks.   Post Code: S11 022

DESCRIPTION AND EXTENT OF THE INSTALLATION Tick boxes as appropriate

Description of installation: *Commercial office*

Extent of installation covered by this Certificate: *Full new installation*

New installation ☑

Addition to an existing installation ☐

Alteration to an existing installation ☐

(Use continuation sheet if necessary)   see continuation sheet No: ....

FOR DESIGN
I/We being the person(s) responsible for the design of the electrical installation (as indicated by my/our signatures below), particulars of which are described above, having exercised reasonable skill and care when carrying out the design hereby CERTIFY that the design work for which I/we have been responsible is to the best of my/our knowledge and belief in accordance with BS 7671:2008, amended to ..2011... (date) except for the departures, if any, detailed as follows:
Details of departures from BS 7671 (Regulations 120.3 and 133.5):
None N/A

The extent of liability of the signatory or the signatories is limited to the work described above as the subject of this Certificate.

For the DESIGN of the installation: **(Where there is mutual responsibility for the design)
Signature: D.Smith... Date: 15/08/2013   Name (IN BLOCK LETTERS): ....D.JONES...   Designer No 1
Signature: .....N/A..... Date: ....   Name (IN BLOCK LETTERS): .....N/A.....   Designer No 2**

FOR CONSTRUCTION
I/We being the person(s) responsible for the construction of the electrical installation (as indicated by my/our signatures below), particulars of which are described above, having exercised reasonable skill and care when carrying out the construction hereby CERTIFY that the construction work for which I/we have been responsible is to the best of my/our knowledge and belief in accordance with BS 7671:2008, amended to ....2011.....(date) except for the departures, if any, detailed as follows:
Details of departures from BS 7671 (Regulations 120.3 and 133.5):
None N/A

The extent of liability of the signatory is limited to the work described above as the subject of this Certificate.

For CONSTRUCTION of the installation:
Signature: T.Smith... Date: 15/08/2013   Name (IN BLOCK LETTERS): ....T.SMITH...

FOR INSPECTION & TESTING
I/We being the person(s) responsible for the inspection & testing of the electrical installation (as indicated by my/our signatures below), particulars of which are described above, having exercised reasonable skill and care when carrying out the inspection & testing hereby CERTIFY that the work for which I/we have been responsible is to the best of my/our knowledge and belief in accordance with BS 7671:2008, amended to ....2011.....(date) except for the departures, if any, detailed as follows:
Details of departures from BS 7671 (Regulations 120.3 and 133.5):
None N/A

The extent of liability of the signatory is limited to the work described above as the subject of this Certificate.

For INSPECTION AND TESTING of the installation:
Signature: G.Wilson... Date: 15/08/2013   Name (IN BLOCK LETTERS): ....G.WILSON...

NEXT INSPECTION
I/We the designer(s), recommend that this installation is further inspected and tested after an interval of not more than ...5.... years/months.

Page 1 of ...4..

Form 2 — Page 2

Form 2
Form No: ...SSSS13........../2

PARTICULARS OF SIGNATORIES TO THE ELECTRICAL INSTALLATION CERTIFICATE

Designer (No 1)
Name: D.Jones   Company: The Electrical Design Partnership
Address: 23 High Street,   Postcode: S10 0JW   Tel No: 01000 444444
Sandtown, Berks.

Designer (No 2) (if applicable)
Name: ....   Company: ....
Address: ....   Postcode: ....   Tel No: ....

Constructor
Name: T.Smith   Company: T.Smith Electrical Installations
Address: Unit 2a, Sandtown Ind.Estate,   Postcode: S12 0XX   Tel No: 01000 999999
Sandtown, Berks.

Inspector
Name: G.Wilson   Company: Wilson and Sons
Address: 11 Outlook Row,   Postcode: S12 0WW   Tel No: 01000 777777
Sandtown, Berks.

SUPPLY CHARACTERISTICS AND EARTHING ARRANGEMENTS Tick boxes and enter details, as appropriate

Earthing arrangements	Number and Type of Live Conductors	Nature of Supply Parameters	Supply Protective Device Characteristics
TN-C ☐	a.c. ☑ d.c. ☐	Nominal voltage, U/U₀(1) ....230... V	Type: BS 88.3
TN-S ☐	1-phase, 2-wire ☐ 2-wire ☑	Nominal frequency, f(1) .....50... Hz	
TN-C-S ☑	1-phase, 3-wire ☐ 3-wire ☐	Prospective fault current, Ipf(2) .141. kA	
TT ☐	2-phase, 3-wire ☐ other ☐	External loop impedance, Ze(2) 0.34Ω	Rated current: 80.. A
	3-phase, 3-wire ☐		
Other sources ☐	3-phase, 4-wire ☐	(Note: (1) by enquiry, (2) by enquiry or by measurement)	
of supply (to be detailed on attached schedules)	Confirmation of supply polarity ☑		

PARTICULARS OF INSTALLATION REFERRED TO IN THE CERTIFICATE Tick boxes and enter details, as appropriate

Means of Earthing	Maximum Demand
Distributor's facility ☑	Maximum demand (load) ....62... kVA / Amps Delete as appropriate

Details of Installation Earth Electrode (where applicable)

Installation earth electrode ☐	Type (e.g. rod(s), tape etc) ...N/A...	Location ...N/A...	Electrode resistance to Earth ...N/A... Ω

Main Protective Conductors

Earthing conductor: material ..Copper..   csa ...16....mm²   Continuity and connection verified ☑
Main protective bonding conductors   material ..Copper..   csa ...16....mm²   Continuity and connection verified ☑
To incoming water and/or gas service ☑   To other elements: ....N/A....

Main Switch or Circuit-breaker

BS, Type and No. of poles .....BS EN 60947.3 (2-pole)..   Current rating .....100.... A   Voltage rating ....400......V
Location .....Office Smith Consumer Unit..   Fuse rating or setting .....N/A.... A
Rated residual operating current IΔn = ...N/A.... mA, and operating time of ...N/A.... ms (at IΔn) (applicable only where an RCD is suitable and is used as a main circuit breaker)

COMMENTS ON EXISTING INSTALLATION (in the case of an addition or alteration see Section 633):
...N/A...

SCHEDULES
The attached Schedules are part of this document and this Certificate is valid only when they are attached to it.
....1.... Schedules of Inspections and ....1.... Schedules of Test Results are attached.
(Enter quantities of schedules attached)

Page .2. of .4.

A competent person is defined in BS 7671 as: 'A person who possesses sufficient technical knowledge, relevant practical skills and experience for the nature of the electrical work undertaken and is able at all times to prevent danger and, where appropriate, injury to him/herself and others'.

Therefore, competent persons must have a sound knowledge and relevant experience of the type of work being undertaken and of the technical requirements of BS 7671. They must also have a sound knowledge of the inspection and testing procedures contained in the Regulations and must use suitable testing equipment.

EICs and MEIWCs must identify who is responsible for the design, construction, inspection and testing, whether this is new work or an alteration or addition to an existing installation.

> **KEY POINT**
>
> An Electrical Installation Condition Report (EICR) will be provided when an inspection and test on an electrical installation has been undertaken in order to highlight any safety shortcomings, defects or deviations from the current version of the Requirements for Electrical Installations (BS 7671).

INTERPRET TESTING DATA

An inspector carrying out the inspection and testing of any electrical installation must have sufficient technical knowledge and experience to carry out the inspection and testing in such a way as to avoid danger to themselves and others. They must have knowledge of relevant technical standards, including BS 7671, and be fully conversant with the required inspection and testing procedures so that they are able to employ suitable test equipment during the test process. The inspector must also have sufficient experience to interpret the results obtained during the test process, being able to take a view and report on the condition of the installation.

It is a requirement that appropriate documentation is retained following testing and inspection, and reports can be produced in any durable medium such as written hard copy or by electronic means. The original copy of the report should be retained in a safe place and be made available to any person inspecting or undertaking work on the electrical installation in the future. If the property is vacated, the report will provide the new owner or occupier with details of the condition of the electrical installation at the time the report was issued.

Unless there are specific values that must be achieved for the installation to be deemed safe, readings such as insulation resistance should be considered relative. Readings obtained on one particular day for a piece of equipment, for example a motor, may not indicate a fault. However, the skill of the electrical technician is to determine what a trend in the readings may represent. For example, readings taken over time may show a trend that indicates failing insulation resistance and the need for some preventative maintenance. Periodic testing is, therefore, the best approach to preventive maintenance of electrical equipment.

PLANNING AND AGREEING PROCEDURES

Before fault diagnosis is carried out, the safe-working arrangements must be discussed and agreed with the client and/or the duty holder (or responsible person for the installation) in a clear, concise and courteous manner. The testing and inspection procedure must be a planned activity as it will almost certainly affect people who work or live in the premises where the installation is being tested. This ensures that everyone who is concerned with the work understands what actions need to be taken, such as:

- which areas of the installation may be subject to disconnection
- anticipated disruption times
- who might be affected by the work
- health and safety requirements for the site
- which area will have restricted access
- whether temporary supplies will be required whilst the fault diagnosis is underway
- reaching an agreement on who has authority for the diagnosis and repair.

It may be that a specific person has responsibility for the safe isolation of a particular section of an installation, so that person should be identified and the isolation arrangements agreed. By entering into dialogue with the client before work commences, the potential for unforeseen events will be minimised and good customer relations will be fostered.

For example, in an office block where the electrical installation is complex and provides supplies to many different tenants located on a number of floors, the safe isolation of a sub-circuit for testing purposes may require a larger portion of the installation to be turned off initially. In order to achieve this with minimal disruption, an agreement must be reached between the competent person tasked to carry out the work and the person responsible for the installations affected. This responsible person could be the office manager, the designated electrical engineer for the site or, in some cases, the landlord of the building.

Everyone involved in the work (for example, client, electrician and those in the workplace) has a responsibility for their own health and safety and that of others who may be affected by the work. Communication between all parties will ensure compliance with the respective health and safety requirements.

Assessment criteria

5.4 Explain how and why relevant people need to be kept informed during completion of fault correction work

KEY POINT

You should appreciate the difference between the duty holder and the responsible person.

The person in control of the danger is the *duty holder*. This person must be competent by formal training and experience and with sufficient knowledge to avoid danger. The level of competence will differ for different items of work.

The person who is designated the *responsible person* has delegated responsibility for certain aspects of a company's operational functions such as fire safety, electrical operational safety or the day-to-day responsibility for controlling any identified risk such as *Legionella* bacteria.

Assessment criteria

5.5 Specify the methods for restoring the condition of building fabric

ASSESSMENT GUIDANCE

Both overhead and underground cables are liable to damage. Overhead lines are susceptible to lightning strikes and inadvertent contact (by fishing poles, masts on dinghys, ladders etc), while underground cables can be damaged during ground excavation works or by sharp tools used in street works and similar operations.

Assessment criteria

5.6 State the methods to ensure the safe disposal of any waste and that the work area is left in a safe and clean condition

SmartScreen Unit 308
Handout 26 and Worksheet 26

RESTORING THE BUILDING FABRIC

There may be a requirement to disturb the fabric or structure of the building and, if this is the case, it is very important for all aspects of the rectification to be discussed with the client. Agreement must be obtained for the work to be undertaken, for the extent of the repair necessary (to brick, block, plaster, concrete, screed, plasterboard and decorations, for example) and for the contractual arrangements (who is paying for the repair). The fabric and structure of the building must always be left in a condition that does not compromise either fire safety or the building's structural performance.

Minor cosmetic repair works such as patch plastering, disturbance to stud walls or decoration are often within the capability of an experienced electrical technician, but you must always recognise your own limitations. Expert advice, such as from a specialist contractor, should be sought if any structural modifications are required.

WASTE DISPOSAL

Another important part of the fault repair process is the safe disposal of waste. This ensures both good customer relations and compliance with the relevant legislation, such as the Waste Electrical and Electronic Equipment (WEEE) Regulations 2006, the Waste (England and Wales) Regulations 2011 and the Control of Asbestos at Work Regulations 2012.

The UK has implemented an EU Directive through the WEEE Regulations 2006, which came into force on 2 January 2007. The regulations apply to all electrical and electronic equipment placed on the market in the UK in any of following 10 product categories:

- large household appliances
- small household appliances
- IT and telecoms equipment
- consumer equipment
- lighting equipment
- electrical and electronic tools
- toys, leisure and sports equipment
- medical devices (except implants and infected products)
- monitoring and control equipment
- automatic dispensers.

The regulations require any 'producer' of such equipment, that is a manufacturer, rebrander or importer of electrical and electronic equipment, to finance the costs of collection and treatment of waste electrical and electronic equipment that arises over a calendar year, in proportion to the amount by weight placed on the market. Producers meet their obligations by registering with an approved *producer compliance scheme*. Through this scheme, producers fund reuse, recovery and recycling of electrical goods at an approved authorised treatment facility (AATF) or approved exporter (AE).

In 2009 there were several amendments made to the UK WEEE Regulations that mainly affect producer compliance schemes, approved authorised treatment facilities and approved exporters.

The UK has also implemented an EU Directive (The Waste Framework Directive), which is the primary European legislation for the management of waste, through a series of regulations dealing with waste. The directive has been revised and these revisions have been implemented in England and Wales through the Waste (England and Wales) Regulations 2011 and ancillary legislation in Wales.

The on-site disposal of waste materials following electrical installation work will be dealt with in a number of ways.

- The packaging material from the electrical fittings and accessories (mainly cardboard) is normally stored and arrangements made for collection, transport and recycling.
- Small amounts of non-recyclable material can be disposed of in the electrical contractor's skip or in the client's skip, if agreement has been reached for that to take place.
- Off-cuts of cable, conduit, trunking, cable tray and general ferrous and non-ferrous materials are often collected for disposal at a metal recycling plant.
- Useable off-cuts of cable, conduit, trunking and cable tray should be returned to stock for future use.

Asbestos

The Control of Asbestos at Work Regulations 2012 affect anyone who owns, occupies, manages or otherwise has responsibilities for the maintenance and repair of buildings that may contain asbestos.

Asbestos materials may be encountered by electricians during the course of their work. Asbestos materials in good condition are usually safe. However, if asbestos fibres become airborne they are very dangerous; this may happen when materials are damaged, due to demolition or remedial works on or in the vicinity of ceiling tiles, asbestos cement roofs, wall sheets, sprayed asbestos coating on steel structures, and lagging.

If asbestos is discovered during electrical installation or remedial work, work must be stopped immediately. Specialist contractors must be engaged to ascertain the condition of the asbestos and to determine any actions necessary for its removal, treatment or retention.

In particular, the disposal of asbestos should only be undertaken by specialist contractors.

Fluorescent tubes

Fluorescent tubes generally contain 94% glass, 4% ferrous and non-ferrous metals and 2% phosphor powder, which itself contains mercury. Fluorescent tubes are classified as hazardous waste in England and Wales and as special waste in Scotland. Preferably, they should be recycled or, if absolutely necessary, taken to specialist disposal sites. They must not be disposed of as general waste.

LEAVING THE INSTALLATION SAFE

Remember that Section 3 of the Health and Safety at Work etc Act and the EAW Regulations require that installations are left in a safe condition. People who have been working on an installation must not leave it in an unsafe condition which could affect contractors, visitors or the general public. For example, where there are accessible live parts due to blanks missing from a consumer unit, suitable temporary barriers should be provided to prevent direct contact with those live parts.

If there is a risk due to a classification C1 defect (as described on page 163) it is *not* sufficient just to make the duty holder aware of the danger when submitting the report. The installation should be made safe, on discovery of the defect, by the person undertaking the inspection and testing. The duty holder or responsible person ordering the report should be advised immediately of the action taken. You should seek agreement for any necessary remedial work to be undertaken straight away or, if that is not practical, you must take other appropriate action, such as switching off and isolating the affected parts of the installation to prevent danger.

ASSESSMENT CHECKLIST

WHAT YOU NOW KNOW/CAN DO

Learning outcome	Assessment criteria	Page number
1 Understand the principles, regulatory requirements and procedures for completing the safe isolation of electrical circuits and complete electrical installations	*The learner can:*	
	1 Specify and undertake the correct procedure for completing the safe isolation of an electrical circuit	144
	2 State the implications of carrying out safe isolations to: ■ other personnel ■ customers/clients ■ public ■ building systems (loss of supply)	146
	3 State the implications of not carrying out safe isolations to: ■ self ■ other personnel ■ customers/clients ■ public ■ building systems (presencc of supply)	148
	4 Identify all Health and Safety requirements which apply when diagnosing and correcting electrical faults in electrotechnical systems and equipment including those which cover: ■ working in accordance with risk assessments/permits to work/method statements ■ safe use of tools and equipment ■ safe and correct use of measuring instruments ■ provision and use of PPE ■ reporting of unsafe situations	148

Learning outcome	Assessment criteria	Page number
2 Understand how to complete the reporting and recording of electrical fault diagnosis and correction work	*The learner can:* **1** State the procedures for reporting and recording information on electrical fault diagnosis and correction work	159
	2 State the procedures for informing relevant persons about information on electrical fault diagnosis and correction work and the completion of relevant documentation	165
	3 Explain why it is important to provide relevant persons with information on fault diagnosis and correction work clearly, courteously and accurately	167
3 Understand how to complete the preparatory work prior to fault diagnosis and correction work	*The learner can:* **1** Specify safe working procedures that should be adopted for completion of fault diagnosis and correction work	168
	2 Interpret and apply the logical stages of fault diagnosis and correction work that should be followed	170
	3 Identify and describe common symptoms of electrical faults	173
	4 State the causes of the following types of fault: ■ high resistance ■ transient voltages ■ insulation failure ■ excess current ■ short circuit ■ open circuit	174
	5 Specify the types of faults and their likely locations in: ■ wiring systems ■ terminations and connections ■ equipment/accessories (switches, luminaries, switchgear and control equipment) ■ instrumentation/metering	176

Learning outcome	Assessment criteria	Page number
	6 State the special precautions that should be taken with regard to the following: ■ lone working ■ hazardous areas ■ fibre-optic cabling ■ electro-static discharge (friction, induction, separation) ■ electronic devices (damage by over voltage) ■ IT equipment (e.g. shutdown, damage) ■ high frequency or capacitive circuits ■ presence of batteries (e.g. lead acid cells, connecting cells)	179
4 Understand the procedures and techniques for diagnosing electrical faults	*The learner can:* 1 State the dangers of electricity in relation to the nature of fault diagnosis work	187
	2 Describe how to identify supply voltages	189
	3 Select the correct test instruments (in accordance with HSE guidance document GS 38) for fault diagnosis work	190
	4 Describe how to confirm test instruments are fit for purpose, functioning correctly and are correctly calibrated	197
	5 State the appropriate documentation that is required for fault diagnosis work and explain how and when it should be completed	197
	6 Explain why carrying out fault diagnosis work can have implications for customers and clients	200
	7 Specify and undertake the procedures for carrying out the following tests and their relationship to fault diagnosis: ■ continuity ■ insulation resistance ■ polarity ■ earth fault loop impedance ■ RCD operation ■ current and voltage measurement ■ phase sequence	201
	8 Identify whether test results are acceptable and state the actions to take where unsatisfactory results are obtained	207

Learning outcome	Assessment criteria	Page number
5 Understand the procedures and techniques for correcting electrical faults	*The learner can:*	
	1 Identify and explain factors which can affect fault correction, repair or replacement	209
	2 Specify the procedures for functional testing and identify tests that can verify fault correction	210
	3 State the appropriate documentation that is required for fault correction work and explain how and when it should be completed	211
	4 Explain how and why relevant people need to be kept informed during completion of fault correction work	215
	5 Specify the methods for restoring the condition of building fabric	216
	6 State the methods to ensure the safe disposal of any waste and that the work area is left in a safe and clean condition	216

ASSESSMENT GUIDANCE

Assignments

- This unit concentrates on fault finding and diagnosis.

- Make sure you have the books and drawing materials you require. These would include BS 7671 IET Wiring Regulations and On-Site Guide. You will need a calculator, not a mobile phone.

- Each assignment is based on a drawing. You must read the notes on the drawing to familiarise yourself with the building detail and the details of the electricity supply and earthing arrangements.

- You should use either a black or a blue pen to complete the assignment. Use other coloured pens to draw any wiring systems or earthing arrangements. Use a straight-edge rule for neatness.

- Make sure you read each question thoroughly. Mistakes are often made as a result of not reading the question fully before answering it.

- The assignment is **not** carried out under exam conditions and you will be allowed sufficient time to complete it.

- You may discuss questions with others but the work submitted must be your own.

- The questions will be supplied beforehand to allow research.

- Make sure you have all the paperwork you need before you begin.

- Arrive on time for the assignment.

- Answer all questions as far as you can.

- When answering questions that require you to describe a process, such as a test procedure, use bullet points to help delineate each stage.

- Plan your approach to the assignment and keep a record of your progress so that you do not have to rush at the end.

Before the assignment

- You will find some questions starting on page 225 to test your knowledge of the learning outcomes.

- Make sure you go over these questions in your own time.

- Spend time on revision in the run-up to the assignment.

Practical assessment

- Task B requires you to carry out fault finding and diagnosis on a rig.
- Make sure you work methodically.
- You may have guidance material with you.
- You must demonstrate competence to your assessor.
- Ensure you check all instruments before use, to ensure they work correctly and are in a safe condition.
- Keep rechecking your instruments for accuracy when carrying out continuity tests; changing leads will necessitate re-nulling or zeroing.
- Ensure you use all necessary Personal Protection Equipment; this should be made available to you.
- Testing is not a race. Doing it right is far better than doing it quickly.
- When you complete a particular stage or test, ensure everything is reconnected or put back before moving on.
- Do not create clutter in your work area. Work tidily and things become easier.
- Ask before you switch off: ask before you switch on.
- Make sure you fill in any paperwork fully. Do not leave any gaps.

Above all else – WORK SAFELY.

Common task

- This task is common to 2357 Units 305, 306, 307 and 308.
- You will be required to answer a number of short answer questions for which you may research the answers. These questions relate to safe isolation. You will then be required to perform safe isolation procedures under the guidance of your assessor.

KNOWLEDGE CHECK

1 Identify three items of PPE that may be required when fault finding in a partially completed block of flats.

2 Information is to be provided to the client regarding activities involved in tracing a fault in a shopping centre. State:

   a) to whom the information should be given

   b) a means of recording the information for future reference.

3 Describe the most likely cause for each of the following faults.

   a) three-phase motor running in reverse

   b) switch-start fluorescent flickering and failing to start

   c) immersion heater overheating

   d) PIR-controlled lamp operating during daylight hours

4 State the precautions to be taken when carrying out:

   a) I R test on lighting circuit containing dimmer switches

   b) I R test on two-way lighting circuit

   c) isolation of installation containing UPS systems

   d) work on or near storage batteries.

5 Name three documents that could be referred to during the fault-finding process.

6 State the possible effect on the following groups during a loss of power in a shopping centre.

   a) fellow workers

   b) shoppers

   c) security staff

   d) shop assistants

7 Identify five factors that would affect the decision to repair or replace defective equipment.

8   State how the following materials should be disposed of at the end of a contract.

   a) cardboard packaging

   b) polystyrene packaging

   c) scrap PVC/copper cable withdrawn during the contract

   d) old transformer oil

9   Identify three outcome codes that could be inserted on an Electrical installation condition report. State the meaning of each code.

10 It is suspected that an RCD is suffering from nuisance tripping. Describe one method of determining the value (in mA) that it is tripping at.

FLOOR PLAN OF A SMALL INDUSTRIAL UNIT SHOWING THE POSITIONS OF ALL PROPOSED ELECTRICAL EQUIPMENT

This floor plan relates to Unit 307 and allows you to appreciate the electrical layout of a small industrial unit. The positions of the incoming water and gas supplies are also highlighted. During your assessments, you will be required to read and understand this type of drawing. The following pages refer to this plan: 33–37 (inspection exercise); 44 (prescribed tests for a small industrial unit); 55–57 (continuity of protective conductors); 58–60 (continuity testing of a lighting circuit); 62–65 (continuity of a ring final circuit); 88–90 (calculation for earth fault loop impedance); 94–95 (earth fault loop impedance for radial socket outlets). This floor plan is also referred to in many activities. Please refer to Appendix 2 for supply characteristics and circuit details.

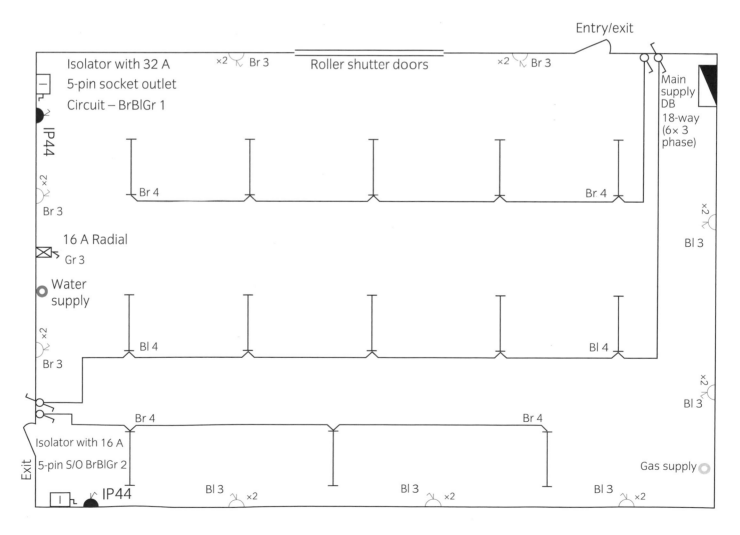

APPENDIX 2: UNIT 307
DISTRIBUTION AND WIRING INFORMATION

Listed below are the supply characteristics and circuit details for the small industrial floor plan shown in Appendix 1, relating to Unit 307.

These supply characteristics and circuit details are supplied to enable you to evaluate and calculate initial verification testing requirements in accordance with BS 7671:2008 (2011) and Guidance Note 3: Inspection and Testing, IET.

By using this information, you could fill in parts of the Electrical Installation Certificate and Generic Schedule of Test Results.

Supply characteristics

Three-phase 230 / 400 V a.c. – 50 HZ

Assessed demand by designer = 48 kVA (approx. 70 A/phase)

Supply authority protective devices = 3 No. 100 A BS 88-3 type 1 fuses

TN-C-S (PME) Z_e at the DB = 0.21 Ω PEFC = 1.09 kA

PSCC = 1.15 kA measured single phase (calculated three-phase [x2] = 2.3 kA)

The circuits

Br/Bl/Gr 1 – C32 A three-phase isolator with socket outlet (5-pin)

Br/Bl/Gr 2 – C16 A three-phase isolator with socket outlet (5-pin)

Br 3 – B32 A ring final circuit for BS 1363 socket outlets

Bl 3 – B32 A radial final circuit for BS 1363 socket outlets

Gr 3 – B16 A radial final circuit for a 3 KW instantaneous water heater

Br 4 – B10 A lighting circuit

Bl 4 – B6 A lighting circuit

Distribution board circuit details

Br/Bl/Gr 1 – C32 A BS EN 60898 3-phase + N Isolator with socket outlet (5-pin)
(all conductors 6 mm^2 including cpc) *measured distance of 15 m*

Br/Bl/Gr 2 – C16 A BS EN 60898 3-phase Isolator with socket outlet (5-pin)
(all conductors 4 mm^2 including cpc) *measured distance of 27 m*

Br 3 – B32 A BS EN 61009 (30 mA) ring final circuit for BS 1363 socket outlets
(2.5 mm^2 line and neutral conductors – cpc 1.5 mm^2) *48 m – total ring (no spurs)*

Bl 3 – B32 A BS EN 60898 radial final circuit for BS 1363 socket outlets
(4 mm^2 line and neutral conductors – cpc 1.5 mm^2) *22 m to final socket outlet*

Gr 3 – B16 A BS EN 60898 radial final circuit for a 3 KW instantaneous water heater – 13 A switch fused spur
(all conductors 2.5 mm^2 including cpc) *measured distance 22 m*

Br 4 – B10 A BS EN 60898 lighting circuit
(all conductors 1.5 mm^2 including cpc) *measured distance 37 m*

Bl 4 – B6 A BS EN 60898 lighting circuit
(all conductors 1.5 mm^2 including cpc) *measured distance 28 m*

Gr 4

Br 5

Bl 5

Gr 5

Br 6

Bl 6

Gr 6

Cable and wiring methods

The wiring is in singles (single-core 70 °C – thermoplastic insulated) cables, installed in a combination of metal and PVC conduit and trunking (individual cpcs supplied for each circuit). All single-phase (single-pole) circuit breakers have a 6 kA rated short-circuit capacity.

ALL three-phase (triple-pole) circuit breakers have a 10 kA rated short-circuit capacity.

GLOSSARY

Breaking capacity The amount of current a protective device can safely disconnect.

C

Competent person A person who possess sufficient knowledge, relevant practical skills and experience for the nature of the electrical work undertaken and is able at all times to prevent danger and, where appropriate, injury to him/herself and others.

Contact resistance Sometimes, a meter may give a reading in which the value does not stabilise or which is quite different from the calculated value. In this situation, reverse the test meter lead connections and re-test. This may give more accurate values.

D

Duty holder The person in control of the danger is the duty holder. This person must be competent by formal training and experience and with sufficient knowledge to avoid danger. The level of competence will differ for different items of work.

E

Extent The amount of inspection and testing. For example, in a third floor flat with a single distribution board and eight circuits, the inspection will be visual only without removing covers and testing, and will involve sample tests at the final point of each circuit.

F

FELV Functional extra-low voltage circuit (requirements can be found in Section 411.7 of BS 7671).

H

Hot work Work of any type which involves actual or potential sources of ignition and which is done in an area where there may be a risk of fire or explosion (for example welding, flame cutting and grinding).

I

Isolation Isolation is the disconnection and separation of the electrical equipment from every source of electrical energy in such a way that this disconnection and separation is secure.

L

Limitation A part of the inspection and test process that cannot be done for operational reasons. For example, the main protective bonding connection to the water system, located in the basement, could not be inspected as a key for the room was not available.

P

PELV Protective extra-low voltage circuit.

Person in control of the premises The person in control of the premises is likely to have physical possession of the premises or is the person who has responsibility for, and control over, the condition of the premises, the activities conducted on those premises and the people allowed to enter.

Personal protective equipment (PPE) All equipment, including clothing for weather protection, worn or held by a person at work, which protects that person from risks to health and safety.

PME Protective multiple earthing.

R

Responsible person The person who is designated as the responsible person has delegated responsibility for certain aspects of a company's operational functions, such as fire safety, electrical operational safety or the day-to-day responsibility for controlling any identified risk such as Legionella bacteria.

S

SELV Separated extra-low voltage circuit.

T

TN-C-S T = source earth (terre).
N = neutral and protective conductor.
C = combined in the distribution system.
S = separated in the installation.

TN-S T = source earth (terre).
N = neutral and protective conductor.
S = separated in the installation.

TT T = source earth (terre).
T = installation earth (terre).

ANSWERS TO ACTIVITIES AND KNOWLEDGE CHECKS

Answers to activities and knowledge checks are given below. Where answers are not given it is because they reflect individual learner responses.

UNIT 307 UNDERSTANDING PRINCIPLES, PRACTICES AND LEGISLATION FOR THE INSPECTION, TESTING, COMMISSIONING AND CERTIFICATION OF ELECTROTECHNICAL SYSTEMS AND EQUIPMENT IN BUILDINGS, STRUCTURE AND THE ENVIRONMENT

Activity answers

Page

3 Regulation 29 provides grounds for defence based on everything practical having been done to prevent such an incident.

5 Make sure you check the casing, as well as the leads and probes.

6 There is always the remote risk of N and E being cross-connected or a voltage appearing on the earth due to a fault. Incorrect polarity or borrowed neutrals can also cause potentially dangerous voltages to appear on the neutral terminals. Although, in theory, these situations should not arise, if you speak to experienced electricians they will recount all kinds of oddities they have encountered. Check and be safe.

7 Gather equipment available at your college or training centre and follow the procedure on pages 4–6, under supervision.

8 If the Line is connected first, the other probe will be live until it is connected to the neutral or earth.

9 1) Shock, burns, fire, arcing and explosion risk.

2) The tester, the public, the client, the customer and other trades.

3) Loss of supply could be very dangerous. Essential supplies must be evaluated by competent persons on-site (usually the electrical engineer).

Page

4) Ensure that the client backs-up the system and switches it off himself/herself.

10 Dangers of working on or near live conductors include risk of electric shock by unintentional contact with conductors through normal body movements and blast or flashover through dropping items such as tools into live parts, or other suitable dangers.

13 At a period determined by the last inspector, depending on the installation condition. The absolute maximum is 10 years; also on change of occupancy (5 years if rented).

15 Design (designer), construction (installer) and test and inspection (test engineer). For small contracts this may be the same person.

20 Someone with sufficient knowledge and experience to carry out the work. This is often interpreted as someone in possession of C&G 2391/2394/5 qualifications. Remember, qualifications are NOT competence.

24 1 a) The Health and Safety at Work etc. Act 1974, The Electricity at Work Regulations 1989, The Management of Health and Safety at Work Regulations 1999.

b) (IET) Wiring Regulations (BS 7671) 17th Edition, IET Guidance Note 3: Inspection and Testing, The IET On-Site Guide (OSG).

2 a) Electrical Installation Certificate, Schedule(s) of Inspections, Generic

Schedule(s) of Test Results, Minor Electrical Installation Works Certificate

b) Electrical Installation Condition Report, Condition Report Inspection Schedule(s), Generic Schedule(s) of Test Results.

3 d) Competent.

25 Alternating current can easily be transformed from one voltage to another by the use of transformers, which allows it to be transmitted over long distances with low losses. Direct current voltage is not easy to change and would result in heavy losses.

26 The type and composition of each circuit, the number and size of conductors, additional protection may need to be considered under certain conditions, the isolating and switching devices, the equipment vulnerable to certain tests.

28 $I_b \leqslant I_n \leqslant I_z$

29 By visual inspection at the terminals when the circuit is isolated.

30 IPX

30 IPXXB.

32 IP4X or IPXXD.

34 Check your answers for this activity on page 37.

38 25 mm^2.

39 It allows the complete isolation of the consumer unit if it is necessary to work on internal connections. Without it some internal parts such as the incoming terminals will still be live.

40 1) Use table 54.7 to find the answer
50 mm^2 ÷ 2 = 25 mm^2.

2) 16 mm^2 tails – 16 mm^2.

3) 4 mm^2 in accordance with this regulation is the minimum.

4) see the equation on page 39.

41 Off position reliably indicated only when contacts fully open. Sufficient contact air gap as designed by the manufacturer.

43 1) Continuity of protective conductors (this test will be carried out during the process of

ring final circuit testing), 2) Ring final circuit test, 3) Insulation resistance test, 4) Polarity test (4 is normally included in 2).

44 They are carried out with the supply isolated – therefore the circuits are dead.

45 Continuity of protective conductors, Insulation resistance, Polarity, Z_s, PFC (verification), Functional test. Tests after polarity do not require any sequence.

48 1) Continuity testing is carried out by using a low resistance ohmmeter and the readings are in ohms (Ω).

2) Insulation resistance testing is carried out by an insulation resistance tester and the readings are in megaohms (MΩ).

49 Check that it is not damaged in any way and the leads are in compliance with GS38.

51 Appendix 3.

51 Properly designed installations provide excellent protection against electric shock, while devices such as RCDs provide additional protection.

52 The conductor would require testing to verify its continuity.

54 1) Test from end to end and record result in milliohms.

2) Use Table B1 of Guidance Note 3 or the IET On-Site Guide Appendix 9, to find the resistance in milliohms per metre and divide into the tested value.

For example, tested value is 460 mΩ and the cable is 18.1 mΩ/m. 460 ÷ 18.1 = 25 m

55 $V = I \times R, I = \dfrac{V}{R}, R = \dfrac{V}{I}$

56 Remove all paint, rust and so on with a file and polish to bright metal with wire wool all the way round the pipe.

57 $R = \dfrac{13 \times 1.83}{1000} = 0.02 \ \Omega$

59 Personnel may trip over it.

61 1) $R_1 + R_2$ = 22 m of 4 mm^2/1.5 mm^2 cable. From the table on page 59 this cable has a resistance of 16.71 mΩ/m:

$$R = \frac{22 \times 16.71}{1000} = 0.37\,\Omega$$

2) *Any* $R_1 + R_2$ = 27 m of cable 4 mm²/4 mm². From Table B1 of Guidance Note 3 or Appendix 9 of the IET On-Site Guide, this cable has a resistance of 9.22 mΩ/m:

$$R = \frac{27 \times 9.22}{1000} = 0.25\,\Omega$$

62 If 2.5 mm² cpcs were used instead of a 1.5 mm² the overall resistance value would be lower.

64 If the L–N and L–E loops are correct then the N–E must be correct.

65 Various test results will apply.

66 $R_t = \dfrac{16 \times 50}{16 + 50} = \dfrac{800}{66} = 12.12\text{M}\Omega$

68 1 ÷ 0.3695 = 2.70 MΩ (see the example on page 67)

69 Switch feed and switch line must be connected together otherwise the wiring beyond the break will not be tested.

70 At the output terminals of the meter or the supply company's double-pole switch if fitted.

73 Ceiling rose.

74 1) Fuses, circuit breakers, switches, 2) circuits for socket outlets, fixed equipment, luminaires, 3) supply tails, distribution boards, consumer units.

76 Use the simple diagram on page 76 and draw in a meter with the probes touching the L terminal and the PE terminal. The meter should read 0.8 Ω.

79 Use the simple diagram on page 79 and draw in a meter with the probes touching the L terminal and the PE terminal. The meter should read 0.35 Ω.

81 Use the simple diagram on page 81 and draw in a meter with the probes touching the L terminal and the PE terminal. The meter should ideally read 200 Ω or less (depending on the supply electrode resistance and the consumer electrode resistance).

84 Memorise and redraw the diagram on page 83.

86 100 mA RCD = 500 Ω, 300 mA RCD = 167 Ω, 500 mA RCD = 100 Ω

90 $Z_s = Z_e + (R_1 + R_2)$, $Z_e = 0.21\ \Omega$

$R_1 + R_2$ = (14.82 mΩ × 22 m) ÷ 1000 = 0.33 Ω (at 20 °C)

Multiply by 1.2 (temperature correction) = 0.40 Ω

Therefore Z_s = 0.21 Ω + 0.40 Ω = 0.61 Ω

Maximum Z_s value allowed for a B16 BSEN 60898 circuit breaker = 2.87 Ω

96 Use Table 41.3 of BS 7671 for this task.

101 Always read the manufacturer's instructions. Some meters can be set up to test between line voltages and other meters could be damaged, causing danger.

103 First press the test button to ensure operation (for new installations). Then the following tests apply. For each of the tests, readings should be taken on both positive (+ve) and negative (−ve) half cycles.

½ × rated tests are carried out to ensure that the RCD is not *too* sensitive.

1× rated tests are carried out to ensure that the RCD is working to the manufactured standard.

5× rated must be used if the RCD is designed for additional protection.

106 Generally a ramp test indicates how sensitive the RCD is to imbalance.

107 At the isolator – it would be dangerous to do it at the motor.

109 All switches, circuit breakers, socket outlets, isolators, luminaires (including two-way switching) to be checked for correct functioning.

110 Answers are dependant on research. Draka and Amtech are two examples, and cable manufacturers also provide simple design packages.

111 Power circuit will be 5% of 400 V = 20 V.

113 3% of 230 V = 6.9 V.

114 The client would not necessarily know which documents need to be handed over on completion. Therefore, it is up to you, as the qualified and competent person, to ensure that the following documents are handed over: full maintenance manual including manufacturer's instructions, Electrical Installation Certificate, Schedule of Inspections, Schedule of Test Results, any

other safety information concerning the electrical installation.

120 Provided the replacement is on a like-for-like basis, a Minor Electrical Installation Works Certificate.

122 Electrical Installation Certificate, Schedule of Inspections, Schedule of Test Results.

Outcome knowledge check answers

Page 141

1	b)	**2**	b)	**3**	a)	**4**	d)	**5**	a)
6	c)	**7**	d)	**8**	b)	**9**	b)	**10**	d)

Knowledge check answers

Page 142

1. a) Identify the ends of each loop and measure
   - i) the line to line resistance

     the neutral to neutral resistance

     the cpc to cpc resistance
   - ii) cross connect L1 to N2 and L2 to N1 and measure / test at each socket / outlet
   - iii) cross connect L1 to cpc 2 and L2 to cpc 1 and measure / test at each socket / outlet

   b) Low resistance ohmmeter

   c) $R = \frac{50 \times 7.41}{1000} = 0.37\,\Omega$ (2)

   d) $R = \frac{0.37 + 0.37}{4} = 0.185\,\Omega$

   e) Individual legs should be identified when the cables are drawn in but if the identification is lost then the resistance between the board and the first socket should be measured. All three conductors should have the same resistance

2. a) Electrical installation certificate. Schedule of test results. Schedule of inspections

   b) Designer, constructor, tester.

   c) Minor works electrical installation certificate

   d) 500v d.c. 1 MΩ

3. a) TT system

   b) PME (TN-C-S) systems should not be exported to outbuildings

   c) Residual current device

   d) 30 mA and 40 milli-seconds

   e) By the use of an earth electrode

UNIT 308 ELECTRICAL INSTALLATIONS: FAULT DIAGNOSIS AND RECTIFICATION

Activity answers

Page

147 A battery.

148 Selection from:
Misaligned switch contacts
Faulty control circuit
Open circuit coil
Loose connections
Damage to containment etc.

156 It limits the torque (turning moment) that can be applied to a screw so it isn't over-tightened.

157 Risk of damage to front of face and ill-fitting helmet being displaced.

162 Condition report inspection schedule
Generic schedule of test results.

163 It is not compulsory to replace the rewireable fuses if they still provide the intended function. Suggest that the board is replaced by a board with circuit breakers/RCDs for added protection and convenience. The code would be C3.

173 A person with sufficient, training, knowledge and experience to carry out the work required.

175 Overload or short circuit causing fuse to blow on supply side.

176 Switch contacts.

178 Fit RCBOs to individual circuits.

Page

181 Ensure that all materials used are conductive to prevent build up of static, see also BS EN 60601.

184 Battery
Generator
Alternative public supply.

186 Electric space heaters, furnaces, immersion heaters or suitable alternatives.

188 The current is very low.

193 Remove from service and send to an approved centre for repair and calibration.

193 As the short circuit occurs the point of the fault separates. This may leave a clear reading.

194 Main earthing conductor.

200 Dustsheet, brush and dustpan, broom, cleaning cloth, vacuum cleaner.

201 1 mA.

204 Divide the ring at approximately the mid-point and test each half, further dividing until the fault is found.

206 The amount of exposed tip for the probes are too long.

218 a) Return to store, b) send for recycling or disposal, c) package up for recycling, d) return to wholesaler or manufacturer for disposal.

Knowledge check answers

Page 225–226

1 a) Overalls, b) hardhat, c) safety glasses or suitable alternatives.

2 a) The client, b) information about the work carried out and any observations about future problems should be recorded by permanent means such as paper forms.

3 a) Phase rotation is incorrect, change over any two phases at the motor terminals, b) faulty tube or starter, c) faulty thermostat and thermal cut-out (where fitted), d) incorrectly set, or faulty sensor in PIR.

4   a) Disconnect dimmer and link out feed and return conductors (switch feed/sw wire), b) operate two-way switches to ensure strappers are properly tested, c) isolate UPS and test as this has a permanent live output, d) risk from discharge gases and short circuiting of terminals by uninsulated spanners etc, make sure room is well ventilated and all tools insulated.

5   a) Circuit charts, b) previous maintenance records, c) manufacturers' data sheets (or suitable alternatives).

6   a) Stop working, loss of income, delays in contract, b) closure of shops, loss of shopping facilities and tills, c) loss of closed circuit TV, d) loss of facilities such as tills, computer systems etc.

7   Age of equipment, availability of spare parts, availability of specialist staff, down-time of plant, loss of production or suitable alternatives.

8   a) Packaged up for recycling, b) can be difficult to recycle but if possible do so, c) packaged up and sent for recycling, d) returned to manufacturer for recycling.

9   C1 danger present, C2 potentially dangerous, C3 improvement recommended.

10   As a basic test the RCD could be tested at 0.5 $I_{\Delta n}$ to see if it trips. If this is not successful, set up a supply with a milliammeter, gradually increasing the current and recording the value at which the RCD trips.

INDEX